Handbook of Crystallography

A.G. Jackson

Handbook of Crystallography
For Electron Microscopists and Others

With 114 Figures

Springer-Verlag
New York Berlin Heidelberg London
Paris Tokyo Hong Kong Barcelona

A.G. Jackson, Ph.D.
Universal Energy Systems, Inc.
Dayton, OH 45432
USA
and
Department of Mechanical Engineering
  and Materials Science
Wright State University
Dayton, OH 45435
USA

Library of Congress Cataloging-in-Publication Data
Jackson, A. G. (Allen G.)
    Handbook of crystallography : for electron microscopists and
others / A.G. Jackson.
      p.   cm.
    Includes bibliographical references and index.
      ISBN-13: 978-1-4612-7776-7
    1. Crystallography — Handbooks, manuals, etc.   I. Title.
QD908.J33   1991
548 — dc20   90-24841

Printed on acid-free paper

9 8 7 6 5 4 3 2 1

ISBN-13: 978-1-4612-7776-7   e-ISBN-13:978-1-4612-3052-6
DOI: 10.1007/978-1-4612-3052-6

Dedicated
to Gene, whose dream was cut short,
to Alice, who found new dreams,
to Drew, who faced death and chose life.

Preface

This book resulted from a series of frustrations. Analytical electron microscopy requires exactly what its name implies: quantitative information to conduct an analysis. The frustrations arose when I started hunting for specific forms of equations in a form understandable to a non-crystallographer, for definitions of subtle concepts related to crystallography, for intelligible interpretations of space group symbols and their significance. What I frequently discovered was that such information was buried in a giant tome and couched in terms familiar to crystallographers but not to electron microscopists in general, or it was located in an old reference not available in my library, or it was found in an out-of-print book, or it was in a Russian book no longer available, etc.

So to minimize the frustrations, I started a notebook containing the details, particularly after I had found forms of equations useful for quick calculations or equations in a form useful for proving, doing, or extending calculations found in a reference. The resulting notebook grew to a respectable size, requiring some organizing of the contents. Finally, the size became large enough, and has proven useful enough, to produce the notebook as a book.

The readers to whom this book is aimed are those microscopists, and others, who are not crystallographers, but who need details about crystal structures in order to interpret images and diffraction effects and patterns present in the electron microscope. Hence, this book is practically oriented. No attempt has been made to make equations mathematically rigorous beyond the rigor needed to accomplish a calculation. Every attempt has been made to insure the mathematical correctness of the equations, because I found correcting mistakes was a major frustration when using an expression from a reference.

This book is not a handbook of electron microscopy. It is a handbook to be used when doing electron microscopy for which crystallographic information plays a fundamental role in the interpretation of the data and images generated. As a minimum my hope is that the book will be a helpful starting reference source for microscopists, since full treatments of the subjects presented would take considerably more pages than I have devoted to them.

The book is divided into 13 chapters. Although the division is somewhat arbitrary, I have grouped items and subjects where possible. Chapter 1 contains mathematics useful in understanding structures, lattices, and the like. Chapter 2 contains specifics for several common crystal systems, bringing together several definitions to be found in the literature. The purpose here is to highlight the fact that different authors use different definitions, which sometimes are not equivalent or are so specialized that their use beyond the one application leaves a reader stranded. Chapter 3 is a brief overview of diffraction pattern analysis. Chapter 4 covers packing and stacking, one of those subjects that lends itself to misinterpretation because of loose use of symbols, particularly the A, B, C site symbols. Chapter 5 describes the seven crystal systems in detail. This chapter contains the equations for each system that one may need to do simple calculations or to refresh one's memory about axes used. Chapter 6 covers transformations of crystal system axes in a practical way. Chapter 7 deals with slip systems. Chapter 8 includes discussions of various projections found to be useful in transmission electron microscopy (TEM). Chapter 9 covers the essentials of symmetry and the wondrous symbolism used in this powerful representation of lattices and structures. Chapter 10 is a brief discussion of convergent beam electron diffraction, especially the use of higher order Laue zone (holz) rings and the parities of the rings. Chapter 11 contains miscellaneous tables that I have found of some value. Chapter 12 is an introduction to icosahedral symmetry and diffraction patterns. The last chapter contains the essentials about dislocations and has tables useful for simple contrast analysis.

Although the list of references is far from complete, they include all those I have found to contain data useful for my needs and in extending my understanding of the fundamental principles underlying crystallography and structures. Of those listed, the following are recommended as good places to go when questions arise:

Barrett, C. S. and T. B Massalski, *Structure of Metals,* Pergamon Press, New York, 1980.
Brown, P. J. and J. B. Forsyth, *The Crystal Structure of Solids,* Edward Arnold Ltd, London, 1979.
Deer, W. A., R. A. Howie, and J. Zussman, *An Introduction to the Rock- Forming Minerals*, Longman Group, Ltd., London, 1980 [available through Halsted Press, Division of John Wiley & Sons, Inc., New York].
Pearson, W. B., *The Crystal Chemistry and Physics of Metals and Alloys*, Wiley- Interscience, New York, 1972.
Vainstein, B. K., A. A. Chernov and L. A. Shuvalov (editors), *Modern Crystallography*, Volumes I-IV, Springer-Verlag, New York, 1979.

It is a pleasure to acknowledge the help and assistance of a number of people. My thanks to Judy Paine for hunting for books in the library when I had given up, to Jeffrey Robbins for his encouragement, to Tom Broderick for helpful discussions concerning slip, to Karen Teal for her encouragement, to Materials Laboratory, and to Systems Research Laboratories for permission to publish, to Dr. C. Suryanarayana for helpful suggestions concerning icosahedral symmetry and comments on the contents, and to my students who have offered suggestions and pointed out errors. I am particularly grateful to the University of Dayton Department of Materials Science for the opportunities to teach from 1984 to 1990. I thank Dr. Dan Miracle for permission to use in Chapter 10 the extended form of the table on diffraction symbols and point groups prior to publication. My thanks also to the production and editorial staff at Springer-Verlag for their constructive suggestions and help.

The following have given permissions to use examples and figures from publications: The Royal Society of London, John Wiley and Sons, ASM International, and the American Physical Society.

Preparation of any book is inevitably a family affair. Hence, my thanks to my spouse, Marti, who played a major role by her expressions of interest and her help in reviewing the early manuscripts, to my daughter, Michelle, for proofing several sections for me, and to my son, Mike, for giving up his tennis table for months.

Any errors present are my responsibility. There are subjects of importance that I have not included, such as matrix representations of symmetry operations, disclinations, and ceramic structures related to minerals. My apologies for such omissions, but the references provided are good starting points for finding more information on these subjects. I will appreciate any comments, thoughts, and suggestions readers may have, complimentary or otherwise.

Preparation of the manuscript was accomplished using the Microsoft Word program on the Macintosh SE. Equations were prepared using the mathematics typesetting codes in Word. Figures were prepared using CricketDraw. The original text was printed on Laserprint paper using a Laserprinter II NT.

Dayton, Ohio   A. G. Jackson
January, 1991

Contents

Preface vii
List of Tables... xv

1. Definitions and Mathematics.. 1
   1.1. Definitions 1
   1.2. Vector Operations 1
       1.2.1. Dot and cross products 1
   1.3. Basis Vectors 2
   1.4. Miller Indices (Plane Indices)... 2
   1.5. Direction Indices... 3
   1.6. Permutation of Indices 4
   1.7. The Direct Lattice 5
       1.7.1. Basis vectors of the direct lattice 5
       1.7.2. Differentiation of the lattice from the structure 5
   1.8. Fundamentals of the Reciprocal Lattice 6

2. Defining Vectors for Various Crystal Systems 8
   2.1. Face-Centered Cubic 8
       2.1.1. Conversion of primitive cell to FCC.. 10
       2.1.2. Other possibilities 10
   2.2. Body-Centered Cubic... 11
   2.3. Hexagonal Close Packed 14
   2.4. End-Centered (Base-Centered).. 15
   2.5. Simple Cubic 16

3. Diffraction Pattern Analysis 17
   3.1. Introduction.. 17
   3.2. Errors in Measurements 17
       3.2.1. Measurements on the emulsion.. 17
       3.2.2. Camera constant equation approximation... 18
       3.2.3. Distortions in patterns 18
       3.2.4. Statistics and error.. 19
   3.3. Analysis of Patterns 20
       3.3.1. Introduction 20
       3.3.2. Indexing conventions 21
       3.3.3. Methods.. 21
           3.3.3.1. Ring patterns 21
           3.3.3.2. Spot patterns 22
   3.4. Structure Factor 24
   3.5. Ratio Tables 25

4. Packing Fraction and Stacking Sequences **28**
    4.1. Packing Fraction 28
        4.1.1. Simple cubic 28
        4.1.2. Face-centered cubic 29
        4.1.3. Body-centered cubic 30
        4.1.4. Hexagonal close packed 30
    4.2. Stacking Sequences 31
        4.2.1. Simple cubic 31
        4.2.2. Body-centered cubic 32
        4.2.3. Face-centered cubic 32
        4.2.4. Hexagonal close packed 33
    4.3. Interstitial Positions 33
        4.3.1. Face-centered cubic 33
        4.3.2. Body-centered cubic 33
        4.3.3. Hexagonal close packed 33

5. Detailed Equations for Various Crystal Systems **38**
    5.1. General Equations Applicable to Any System 38
    5.2. Cubic System 41
    5.3. Tetragonal System 44
    5.4. Orthorhombic System 48
    5.5. Monoclinic System (b axis unique) 52
    5.6. Triclinic System 56
    5.7. Trigonal System 57
        5.7.1. Basis vectors 57
        5.7.2. Various equations 58
    5.8. Hexagonal System 64

6. Conversion Formulas **75**
    6.1. Introduction 75
    6.2. BCC to Orthorhombic 75
    6.3. hcp to Orthorhombic 76
    6.4. BCC to hcp 77
    6.5. FCT to BCT and BCT to FCT Transformation 77
        6.5.1. FCT to BCT 77
        6.5.2. BCT to FCT 78
    6.6. Monoclinic Nonprimitive to Primitive Transformation 78
    6.7. Rhombohedral to HCP 79
    6.8. Some Orientation Relationships 80
    6.9. Some Ordered Structures 80

7. Slip Systems **83**
    7.1. Face-Centered Cubic 83
    7.2. Body-Centered Cubic 83
    7.3. Hexagonal Close Packed 83
    7.4. Miscellaneous Definitions 86

8. Projections 89
    8.1. Introduction 89
    8.2. Direct Lattice Projections 89
    8.3. Reciprocal Space Projections 90
    8.4. The Stereographic Projection 92
    8.5. Grid Projections 96

9. Structure Symbols 100
    9.1. Crystal Designations 100
        9.1.1. Introduction 100
        9.1.2. Equivalent points 100
    9.2. Strukturbericht Symbols 104
    9.3. Pearson Symbols 104
    9.4. Symmetry Symbols 105
        9.4.1. Operational definitions 105
        9.4.2. Macroscopic symmetry elements 105
        9.4.3. Space group symbols 105
        9.4.4. International symbols 106
        9.4.5. Schoenflies symbol 106
        9.4.6. International or Hermann-Mauguin symbol 106
        9.4.7. Jagodzinski-Wyckoff notation 107
        9.4.8. Ramsdell notation 108
        9.4.9. Zhdanov notation 108

10. Convergent Beam Electron Diffraction 127
    10.1. Introduction 127
    10.2. Problems with Obtaining and Interpreting the Patterns 128
        10.2.1. Instrument and specimen related problems 128
        10.2.2. Recording the pattern 128
        10.2.3. Interpretation 128
    10.3. Zero and Higher Order Laue Zones 129
    10.4. Lattice Parameter Along the Zone Axis 130
    10.5. Higher Order Laue Zone Lines in Diffraction Discs 131
    10.6. Symmetry Identification 135
        10.6.1. Some definitions 135
        10.6.2. Point and space group determination 136
    10.7. Thickness Measurement Using Higher Order Laue Zone Lines
          in a Diffraction Disc 136
    10.8. Indexing Holz Patterns 137
    10.9. Construction of the Holz Pattern and
          Identification of Planes in the Holz Ring 140
    10.10. Rings in Convergent Beam Diffraction 142
    10.11. Interpretation of hcp CBED Ring Patterns 146

11. Miscellaneous Tables and Data 162
    11.1. Mendeleev Number and Chemical Scale 162
    11.2. Machlin Classification of Some Intermetallics 162
    11.3. Schlafli Symbols 163
    11.4. Fourier Series and Transforms 167

11.4.1. Introduction 167
11.4.2. Fourier series 167
11.4.3. Fourier transforms 168
11.4.4. Crystal structures and Fourier transforms 170

12. Icosahedral Structures and Patterns **175**
12.1. Definitions 175
12.1.1. Golden mean 175
12.1.2. Icosahedron 175
12.2. Axes 176
12.3. Simple Projection Examples 179
12.4. Diffraction Patterns 181

13. Dislocations **183**
13.1. Definitions 183
13.2. Image Contrast of Dislocations 184
13.3. Analysis of Burger's Vector 185
13.4. Thompson Tetrahedron for Face Centered Cubic 186
13.5. Partials 186
13.6. Twins 189

References **198**

Index **201**

List of Tables

Table 3.1.   Some Values of Structure Factors $| \text{F} |^2$ 25
Table 3.2.   Ratio Tables for FACE-CENTERED CUBIC Lattice 26
Table 3.3.   Ratio Tables for BODY-CENTERED CUBIC Lattice 26
Table 3.4.   Ratio Tables for Simple Cubic Lattice 27
Table 3.5.   Plane Ratios for the Tetragonal System for (hk0) Planes................... 27
Table 3.6.   Ratio Tables for Hexagonal Close Packed Lattice for l = 0 27

Table 4.1.   Unit Cell Parameters for SC, FCC, BCC (after Kittel, [1967, p. 18, table 3])... 30

Table 5.1.   Trigonal Indices in Rhombohedral and Hexagonal Bases 62
Table 5.2.   Indices for Various Hexagonal Indexing Systems 73

Table 6.1.   Special Transformations for BCC, hcp, and Orthorhombic Lattices .. 75
Table 6.2.   Disordered to Ordered Transformations... 81

Table 7.1.   Number of Independent Slip Systems for hcp.................................... 87
Table 7.2.   Number of Independent Slip Systems for FCC, BCC, and hcp 87
Table 7.3.   Burgers Vectors of Dislocations in hcp Structures............................. 88

Table 9.1a. Characteristics of The Seven Crystal Systems 109
Table 9.1b. Characteristics of the Crystal Systems 110
Table 9.2.   Glide Elements 111
Table 9.3.   Definitions of the Strukturbericht Symbols.. 112
Table 9.4.   Definitions of the Pearson Symbols 112
Table 9.5.   The 32 Point Groups and Their Symbols 112
Table 9.6.   Various Symbols Used for Point Groups 113
Table 9.7.   Strukturbericht Symbols, Archtypes, Pearson Symbol and Space Group Arranged by Strukturbericht Symbol...................................... 115
Table 9.8.   Strukturbericht Symbols, Archtypes, Pearson Symbol and Space Group Arranged by Archtype. 118
Table 9.9.   Strukturbericht Symbols, Archtypes, Pearson Symbol and Space Group Arranged by Pearson Symbol 121
Table 9.10. Strukturbericht Symbols, Archtypes, Pearson Symbol and Space Group Arranged by Space Group.. 124

Table 10.1. Sums of a Zolz and a Folz Vector $(11\bar{1})$ to Produce g Vectors in the FCC Folz.. 138
Table 10.2. Values of h,k, and l for which $h^2 + k^2 + l^2$ is near $G^2a^2 = 165.73$ (for a = 0.3 nm. H = $1/3\sqrt{2}$ = 0.2357 nm^{-1}; G^2 = 1.841 nm^{-2})... 142
Table 10.3. Experimentally Observed and Calculated Values of H for Various hcp Planes of TiB$_2$.. 148
Table 10.4. Expected Values of H$_{exp}$ in Terms of H$_{theor}$ for Various hcp Planes in (4, 4) and (3, 3) Notation........................... 148

Table 10.5. Forms for H for the Seven Crystal Systems 149

Table 10.6. Diffraction Groups, Zones, Point Groups and Crystal Systems
Arranged by Diffraction Group.. 150

Table 10.7. Diffraction Groups, Zones, Point Groups and Crystal Systems
Arranged by Point Group. 152

Table 10.8. Diffraction Groups, Zones, Point Groups and Crystal Systems
Arranged by Crystal System. 154

Table 10.9. Diffraction Groups, Zones, Point Groups and Crystal Systems
Arranged by Zone... 156

Table 10.10. Matrix Relating the Diffraction Groups (vertical)
to the Point Groups (horizontal) 158

Table 10.11. Symmetry Symbols for CBED Discs for the Point Groups.............. 159

Table 10.12. Expressions for u* and for H/u* Used in the Calculation
of Translation Vectors... 161

Table 11.1. Mendeleev Number M and Chemical Scale (χ) 171

Table 11.2. Lattice Parameters of Some Intermetallic Compounds by
Strukturbericht Symbol and Machlin Classification 172

Table 11.3. Some Common Fourier Transforms..................................... 174

Table 12.1. A Listing of Peaks for the Icosahedral Quasicrystal
Having $|n_\parallel|$ <25 and $n_\perp$< 3.4... 179

Table 13.1. Stair-rod Reactions with Values of b^2 for Shockley Partials............ 191

Table 13.2. Major Thompson Tetrahedron Vectors and Associated Directions... 191

Table 13.3. Burgers Vectors of Stable Dislocations in
FCC, BCC, hcp, Diamond Cubic and NaCl.................................... 191

Table 13.4. g•b for FCC Lattice 192

Table 13.5. FCC Imperfect Dislocations; Values of g•b
for Various Planes and for b = 1/6<110>..................................... 193

Table 13.6. FCC Imperfect Dislocations.; Values of g•b
for Various Planes and for b = 1/3<111>..................................... 193

Table 13.7. FCC Imperfect Dislocations; Values of g•b
for Various Planes and for b = 1/6<112>..................................... 194

Table 13.8. g •b Values for BCC Perfect Dislocations..................................... 195

Table 13.9. Details of Various Vectors Used in hcp Vector Notation
for Dislocations 196

Table 13.10. Hcp Perfect Dislocations; Values of g•b for Various Planes
and for b = 1/3<$\overline{1}\overline{1}$23> and 1/3[0003] 197

Table 13.11. Hcp Perfect Dislocations; Values of g•b
for Various Planes and for b = 1/3<$\overline{1}\overline{1}$20>. 197

1
Definitions and Mathematics

1.1. Definitions

crystal ≡ a material in which the atoms are arranged in a translationally periodic array in three dimensions or which are arranged in rotationally periodic arrays in three dimensions [Cullity , 1978].

lattice ≡ array of points in three-dimensional space; each point displays identical symmetry.

crystallographic point ≡ a point in a lattice that displays the symmetry of the lattice. The crystallographic point is not the same as a mathematical point, which has zero dimensions and has no symmetry associated with it [Vainstein, 1979].

unit cell ≡ a volume characterized by vectors **a**,**b**,**c** which are taken from a common origin and are all not coplanar. Repetition of this cell generates the lattice.

a_1, a_2, a_3 ≡ alternate notation for axes **a**, **b**, **c**.

crystallographic axes ≡ **a**, **b**, **c** of the unit cell.

lattice constant(s) ≡ lattice parameters; the repeat distances along each of the crystallographic axes; magnitudes of the vectors **a**, **b**, **c**, and angles between axes.

basis vectors ≡ unit vectors in terms of which **a**, **b**, **c** are defined.

orthogonal basis ≡ set of vectors that satisfy $e_i \cdot e_j = \delta_{ij}$.

basis ≡ **motif** ≡ group of atoms associated with a lattice point.

primitive cell ≡ **simple cell** ≡ **P** ≡ lattice unit cell with one lattice point per cell, or the minimum number of points possible in the lattice unit cell.

Laue class ≡ symbol expressing the point symmetry of the crystal system as observed in the diffraction pattern; the symmetry of the reciprocal lattice times the intensity of the (hkl) plane.

Laue zone ≡ those reciprocal lattice points that lie in planes parallel to the first reciprocal lattice plane that intercepts the Ewald sphere; $g \cdot u = N$, where g = reciprocal lattice vector, u = direction vector defining the diffraction zone axis, N = integer = number of the Laue zone.

zone axis ≡ the axis formed by those reciprocal lattice vectors which satisfy $g \cdot u = 0$; the axis is parallel to the direction vector **u**.

1.2. Vector Operations

1.2.1. Dot and cross products

For a triad of vector axes (**a**, **b**, **c**) defined in terms of an orthonormal reference coordinate system (e_1, e_2, e_3), the two most frequently used vector operations in electron diffraction crystallography) are :

Dot product **a** · **b**, where

$$a = a_1 e_1 + a_2 e_2 + a_3 e_3, \qquad (1.1)$$
$$b = b_1 e_1 + b_2 e_2 + b_3 e_3, \qquad (1.2)$$
$$c = c_1 e_1 + c_2 e_2 + c_3 e_3, \qquad (1.3)$$

is given by

$$a \cdot b = a_1 b_1 + a_2 b_2 + a_3 b_3 = b \cdot a$$
$$= |a| |b| \cos(a, b). \qquad (1.4)$$

Cross product a x b: The sense of the cross product is always right handed in a right-handed coordinate system. The cross product can be defined as

$$a \times b = \begin{bmatrix} e_1 & e_2 & e_3 \\ a_1 & a_2 & a_3 \\ b_1 & b_2 & b_3 \end{bmatrix}$$

$$= e_1 (a_2 b_3 - b_2 a_3) - e_2 (a_1 b_3 - b_1 a_3)$$
$$+ e_3 (a_1 b_2 - b_1 a_2)$$

$$= c' = |a| |b| \sin(a, b). \qquad (1.5)$$

Note the following equations:

$$a \times a = b \times b = 0, \quad c' \cdot a = c' \cdot b = 0. \qquad (1.6)$$

The mixed product is used to define the volume of a unit cell as

$$V = a \cdot (b \times c). \qquad (1.7)$$

1.3. Basis Vectors

For a set of vectors (e_1, e_2, e_3) that are orthonormal, i.e., $e_1 \cdot e_2 = e_2 \cdot e_3 = e_3 \cdot e_1 = 0$ and $e_i \cdot e_i = 1$ for $i = 1,2,3$, the basis coordinate system is defined to be right handed if $e_1 \times e_2 = e_3$, $e_3 \times e_1 = e_2$, and $e_2 \times e_3 = e_1$. The unit vectors i, j, k are frequently used also, and $i = e_1, j = e_2, k = e_3$. See Figure 1.1.

1.4. Miller Indices (Plane Indices)

The reference coordinate system is shown in Figure 1.2, where the vectors defining the unit cell are a_1, a_2, a_3. A plane is defined by the intersection with each axis as shown. The distance from the origin to each intersection can be written as a fraction of some distance along each axis. For the case

$$a_1 = a e_1, \qquad (1.8)$$
$$a_2 = b e_2, \qquad (1.9)$$
$$a_3 = c e_3, \qquad (1.10)$$

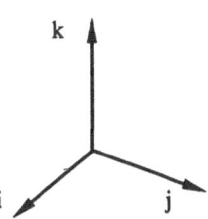

Figure 1.1. Right-handed coordinate system showing the unit vectors **i**, **j**, **k**.

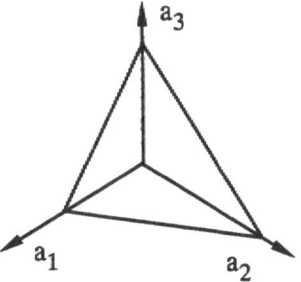

Figure 1.2. Intercepts of the axes which define the plane and the Miller indices.

where (e_1, e_2, e_3) are defined such that $|e_1| = |e_2| = |e_3| = 1$ and $e_i \cdot e_j = \delta_{ij}$, and $|a_1| = a$, $|a_2| = b$, $|a_3| = c$ (a, b, c not necessarily equal), the intercept of the plane with each axis can be written

$$\left(\frac{a}{h} e_1, \frac{b}{k} e_2, \frac{c}{l} e_3 \right). \tag{1.11}$$

[Note: $\delta_{ij} = 1$ for i = j and 0 for i ≠ j; i, j = 1, 2, 3.]

The (hkl) are called the Miller indices of the plane, and they are used to label the plane with the restriction that h,k,l must each be an integer. Notation in use is (hkl) for a specific set of integers, and {hkl} for a type of plane. See the permutation of indices in Section 1.6.

1.5. Direction Indices

The Miller indices label planes using a set of integers (hkl). To label directions in the direct lattice of the crystal, a different set of integers is used. The directions in the direct lattice can be represented by the vector from an origin to the atom position, which is defined, in terms of the lattice basis vectors and integers u, v, w, as

$$\mathbf{u} = u\, a_1 + v\, a_2 + w\, a_3 = <u,v,w> \tag{1.12a}$$

for a nonspecific direction, and

$$\mathbf{u} = [uvw] \tag{1.12b}$$

for a specific set of values of u, v, w.

From the definitions of Miller indices, one has

$$A = \frac{a}{h} e_1, \quad B = \frac{b}{k} e_2, \quad C = \frac{c}{l} e_3. \tag{1.13}$$

$\mathbf{D}$ is the vector from the origin to the plane, as shown in Figure 1.3. The vectors lying in the plane are perpendicular to $\mathbf{D}$ and also to $\mathbf{u}$. By vector addition one obtains

$$\mathbf{D} - \frac{a}{h}\mathbf{e}_1 = A \cdot \mathbf{D}, \quad \mathbf{D} - \frac{b}{k}\mathbf{e}_2 = B \cdot \mathbf{D}, \quad \mathbf{D} - \frac{c}{l}\mathbf{e}_3 = C \cdot \mathbf{D}. \qquad (1.14)$$

Taking the dot product with $\mathbf{D}$ and setting the result equal to zero gives

$$\frac{a}{h}\mathbf{e}_1 \cdot \mathbf{D} = |\mathbf{D}|^2, \quad \frac{b}{k}\mathbf{e}_2 \cdot \mathbf{D} = |\mathbf{D}|^2, \quad \frac{c}{l}\mathbf{e}_3 \cdot \mathbf{D} = |\mathbf{D}|^2. \qquad (1.15)$$

Eliminating $|\mathbf{D}|^2$ from these equations gives the set of relationships between the plane indices (h,k,l) and the direction indices [u,v,w]:

$$\frac{a^2 u}{h} = \frac{b^2 v}{k} = \frac{c^2 v}{l} \qquad (1.16)$$

for orthonormal axes, and h, k, l, u, v, w integer.

For nonorthonormal axes, see monoclinic, triclinic, trigonal or hcp sections for specific relationships among the plane and direction indices.

1.6. Permutation of Indices

The permutations permitted define the planes that are equivalent in terms of the magnitude of the repeat separation in direct or reciprocal space. The permutations do not define equal planes, because there is a nonzero angle between any two planes derived by permuting the indices.

Note that in each system {hkl} is equivalent to {-h,-k,-l} and that <uvw> is equivalent to <-u,-v,-w>. For definitions of crystal systems see Chapter 5.

Permutation of indices					
	(h,k,l)	<u,v,w>		(h,k,l)	<u,v,w>
cubic	all indices permute		tetragonal	h,k	u,v
orthorhombic	none	none	monoclinic	none	none
triclinic	none	none	rhombohedral	none	none
hexagonal:					
	3 axis, 3 index		4 axis, 4 index	orthohexagonal	
planes	h,k		h, k; i = - (h + k)	none	
directions	u,v		u, v; t = - (u + v)	none	

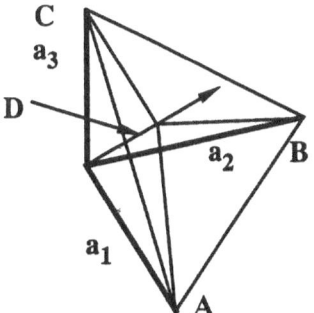

Figure 1.3. Plane in direct space intersecting the coordinate axes, labeled as in the equations.

1.7. The Direct Lattice

1.7.1. Basis vectors of the direct lattice

The lattice defined for a given crystal structure is the direct lattice (also called crystal direct lattice, space lattice, direct space lattice) defined. in terms of a set of basis vectors with properties as described in Section 1.3. Usual notation is a_1, a_2, a_3 (see Figure 1.4).The directions of these vectors coincide with the coordinates chosen for the particular crystal system, and the magnitude of each is chosen to be proportional to the translational repeat distance along the appropriate axis. Hence, any point in the direct lattice can be represented by the set (ua_1, va_2, wa_3), where u,v,w are integers.

1.7.2. Differentiation of the lattice from the structure

The lattice is the construction to which the symmetry is referred. In the simple case of most elemental lattices, the atom is considered to be located at the lattice point. A more general case (and more realistic) is that in which the atom sites and the lattice points do not coincide. In this case each lattice point is associated with a motif consisting of more than one atom in which the atom sites are described by additional position vectors

$$r_{in} = x_{in} a_1 + y_{in} a_2 + z_{in} a_3, \qquad (1.17)$$

where the n refers to the atom species and the i refers to the ith atom of species n.

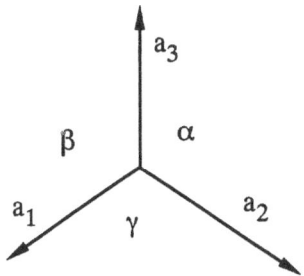

Figure 1.4. Direct lattice coordinate system.

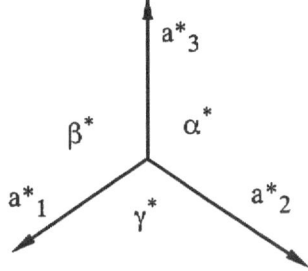

Figure 1.5. Reciprocal lattice coordinate system.

1.8. Fundamentals of the Reciprocal Lattice

For a given direct lattice defined by vectors a_1, a_2, a_3 and integers u,v,w, a vector in this lattice can be written as

$$R = u \, a_1 + v \, a_2 + w \, a_3. \tag{1.18}$$

The a_i are not necessarily orthonormal, i.e., not necessarily a Cartesian coordinate system.

Construct a different lattice, as shown in Figure 1.5, using vectors a^*_1, a^*_2, a^*_3, and integers h, k, l such that

$$R^* = h \, a^*_1 + k \, a^*_2 + l \, a^*_3. \tag{1.19}$$

The set a^*_1, a^*_2, a^*_3 is a reciprocal lattice to the set a_1, a_2, a_3, if the following equalities hold:

I. $\quad a^*_1 = \dfrac{(a_2 \times a_3)}{V}$, $\qquad$ (1.20)

$\quad a^*_2 = \dfrac{(a_3 \times a_1)}{V}$, $\qquad$ (1.21)

$\quad a^*_3 = \dfrac{(a_1 \times a_2)}{V}$; $\qquad$ (1.22)

II. $V = a_1 \cdot a_2 \times a_3 = $ volume of the unit cell; $\qquad$ (1.23)

III. $a_i \cdot a^*_j = \delta_{ij}$. $\qquad$ (1.24)

Note that $a_i \cdot a_j$ and $a^*_i \cdot a^*_j$ may not be zero when $i \neq j$. The units in reciprocal space are (length)$^{-1}$.

The interplanar spacing in direct space is equal to the distance between equivalent planes in a set {h,k,l}, i.e.,

$$1 / | R^* | = d_{hkl}, \quad \text{units = length.} \tag{1.25}$$

For the case suitable for any triad of vectors at known angles to each other,

$$\frac{1}{(d_{hkl})^2} = h^2 a^*_1{}^2 + k^2 a^*_2{}^2 + l^2 a^*_3{}^2 + 2 \, h \, k \, | a^*_1 | \, | a^*_2 | \cos(\gamma^*)$$
$$+ 2 \, k \, l \, | a^*_2 | \, | a^*_3 | \cos(\alpha^*) + 2 \, l \, h \, | a^*_3 | \, | a^*_1 | \cos(\beta^*). \tag{1.26}$$

The angles α^*, β^*, γ^* are the angles between axes in reciprocal space and are not necessarily equal to the corresponding angles in direct space.

Expressions for the reciprocal lattice angles in terms of the direct lattice angles are

$$\cos(\alpha^*) = \frac{\cos \beta \cos \gamma - \cos \alpha}{\sin \gamma \, \sin \beta}, \tag{1.27}$$

$$\cos(\beta^*) = \frac{\cos \alpha \cos \gamma - \cos \beta}{\sin \alpha \, \sin \gamma}, \tag{1.28}$$

$$\cos{(\gamma^*)} = \frac{\cos\alpha\cos\beta - \cos\gamma}{\sin\alpha\sin\beta}. \tag{1.29}$$

The volume in general is given by

$$V = a\,b\,c\,[\,1 - \cos^2\alpha - \cos^2\beta - \cos^2\gamma + 2\cos\alpha\cos\beta\cos\gamma\,]^{1/2}, \tag{1.30}$$

where a, b, c̄ are the lattice parameters along the appropriate axes.

The angle between two planes with indices $(h_1 k_1 l_1)$ and $(h_2 k_2 l_2)$ is in general given by

$$\begin{aligned}
\cos{(\varphi_{1\,2})} = [h_1\,h_2\,&{a^*_1}^2 + k_1\,k_2\,{a^*_2}^2 + l_1\,l_2\,{a^*_3}^2 \\
&+ (h_1\,k_2 + h_2\,k_1)\,|\,a^*_1\,|\,|\,a^*_2\,|\cos\gamma^* \\
&+ (k_1\,l_2 + k_2\,l_1)\,|\,a^*_2\,|\,|\,a^*_3\,|\cos\alpha^* \\
&+ (l_1\,h_2 + l_2\,h_1)\,|\,a^*_3\,|\,|\,a^*_1\,|\cos\beta^*] \\
&\times d(h_1 k_1 l_1)\,d(h_2 k_2 l_2)\,.
\end{aligned} \tag{1.31}$$

Note that, in some physics texts, there is a factor of 2π included in the definitions in order to relate the momentum wavevector and the reciprocal lattice. Hence, the definitions are modified as indicated below:

$$\mathbf{R}^* = \mathbf{K}/2\,\pi\,, \tag{1.32}$$

or

$$a_i \bullet a^*_j = 2\,\pi\,\delta_{ij} \quad (i,j = 1, 2, 3), \tag{1.33}$$

or

$$a^*_i = \frac{2\,\pi\,a_j\times a_k}{V} \quad (i, j, k = 1, 2, 3). \tag{1.34}$$

2
Defining Vectors
for Various Crystal Systems

2.1. Face-Centered Cubic

There are several definitions for the basis vectors for face-centered cubic (FCC) structures and lattices. The most common uses the nonprimitive cube with axes **A**, **B**, **C** as shown in Figure 2.1. Conceptually, this set is the easiest to grasp, but the primitive set is essential for finding the allowed reflections.

$$\mathbf{a} = \frac{a}{2}(\mathbf{i}+\mathbf{j}), \qquad |\mathbf{a}| = \frac{\sqrt{2}\,a}{2}, \tag{2.1}$$

$$\mathbf{b} = \frac{a}{2}(-\mathbf{i}+\mathbf{j}), \qquad |\mathbf{b}| = \frac{\sqrt{2}\,a}{2}, \tag{2.2}$$

$$\mathbf{c} = \frac{a}{2}(-\mathbf{j}+\mathbf{k}), \qquad |\mathbf{c}| = \frac{\sqrt{2}\,a}{2}, \tag{2.3}$$

$$\theta_{a,b} = 90°, \qquad \theta_{a,c} = \theta_{b,c} = 120°, \tag{2.4}$$

and

$$\mathbf{a} \cdot \mathbf{b} = 0, \qquad \mathbf{b} \cdot \mathbf{c} = \mathbf{c} \cdot \mathbf{a} = -\frac{a^2}{4}. \tag{2.5}$$

Note that the **a**, **b**, **c** vectors are the vectors connecting the *closest* atom centers.
Transform from the primitive vectors to FCC vectors using

$$
\begin{aligned}
&\mathbf{A} = \mathbf{a} - \mathbf{b} = a\,\mathbf{i}, & |\mathbf{A}| = a, & \quad(2.6a)\\
&\mathbf{B} = \mathbf{a} + \mathbf{b} = a\,\mathbf{j}, & |\mathbf{B}| = a, & \quad(2.6b)\\
&\mathbf{C} = \mathbf{a} + \mathbf{b} + 2\,\mathbf{c} = a\,\mathbf{k}, & |\mathbf{C}| = a, & \quad(2.6c)\\
&\theta_{A,B} = \theta_{B,C} = \theta_{C,A} = 90°. & & \quad(2.6d)
\end{aligned}
$$

An alternate set of vectors for defining the FCC structure is, according to Ashcroft and Mermin [1976], given by

$$\mathbf{a} = \frac{a}{2}(\mathbf{j}+\mathbf{k}), \qquad |\mathbf{a}| = \frac{\sqrt{2}\,a}{2}, \tag{2.7a}$$

$$\mathbf{b} = \frac{a}{2}(\mathbf{i}+\mathbf{k}), \qquad |\mathbf{b}| = \frac{\sqrt{2}\,a}{2}, \tag{2.7b}$$

8

$$c = \frac{a}{2}(\mathbf{i} + \mathbf{j}), \qquad |c| = \frac{\sqrt{2}\,a}{2}, \tag{2.7c}$$

$$\theta_{1,2} = \theta_{2,3} = \theta_{3,1} = 60°, \tag{2.7d}$$

$$\mathbf{a} \cdot \mathbf{b} = \mathbf{b} \cdot \mathbf{c} = \mathbf{c} \cdot \mathbf{a} = \frac{a^2}{4}, \tag{2.7e}$$

where a is the cubic lattice parameter, i.e., a = the separation of the lattice points along the principal axes of a primitive cube. The angle between **c** and the plane formed by $\mathbf{a} + \mathbf{b}$ is 45°. The cube diagonal is $|\mathbf{a} + \mathbf{b} + \mathbf{c}| = \sqrt{3}\,a$. This set of vectors defines the *rhombohedral* unit cell for FCC structures. The primitive vectors are not orthogonal to each other. See Figure 2.2.

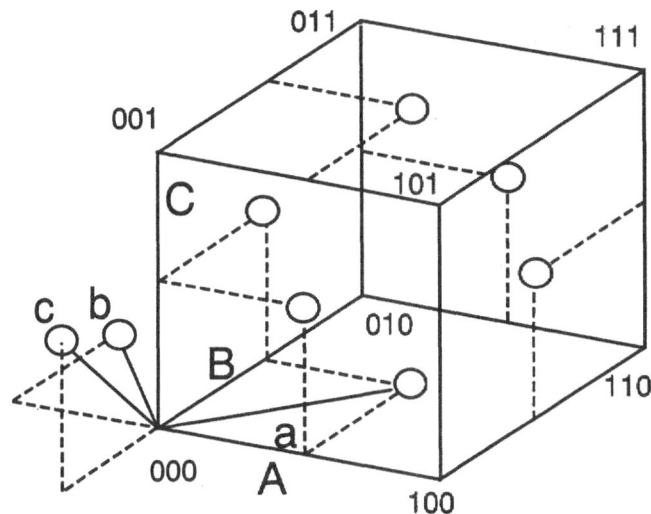

Figure 2.1. Primitive vectors **a**, **b**, **c** and FCC vectors **A**, **B**, **C**.

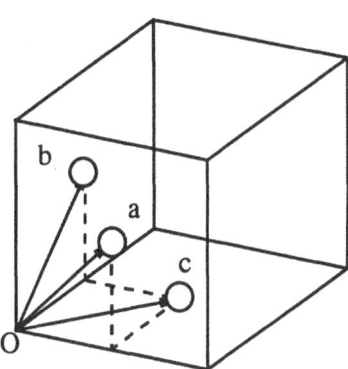

Figure 2.2. Rhombohedral defining axes for FCC.

Let a' = the separation of lattice points along a cube face diagonal of a primitive cube. Then

$$a' = \sqrt{2}\, a, \tag{2.8}$$

where a is the cubic lattice parameter, and the "a" for the rhombohedral definition is

$$a_{rhomb} = \frac{a}{2}, \tag{2.9}$$

which is half the distance along the face diagonal of the cubic lattice.

2.1.1. Conversion of primitive cell to FCC

The FCC basis vectors are **A, B, C**. From the rhombohedral primitive cell defined in the previous section (second definition), the transformation to the FCC basis is given as

$$A = a - b + c = a\,j, \tag{2.10a}$$
$$B = a + b - c = a\,k, \tag{2.10b}$$
$$C = -a + b + c = a\,i. \tag{2.10c}$$

The indices of planes in the two bases are related as

$$H = h - k + l, \tag{2.11a}$$
$$K = h + k - l, \tag{2.11b}$$
$$L = -h + k + l. \tag{2.11c}$$

From this last set of equations, we can identify the plane indices that are allowed and those that are extincted. The allowed planes are determined by taking pairs of indices and summing them; thus

$$H + K = 2\,h, \tag{2.12a}$$
$$K + L = 2\,k, \tag{2.12b}$$
$$L + H = 2\,l. \tag{2.12c}$$

Because each sum of pairs is an even integer, H, K, L must be either all even or all odd, since odd + odd = even + even = even satisfies the sum rules above. Values of H, K, L that do not satisfy these rules are extincted. Hence, indices with mixed parities are forbidden. The same result is obtained using the first definition of primitive vectors.

2.1.2. Other possibilities

The FCC cell can be obtained from a body centered tetragonal (BCT) unit cell, which has basis vectors

$$a_T = a_T\, i, \quad b_T = a_T\, j, \quad c_T = \sqrt{2}\, a_T\, k. \tag{2.13}$$

The transformation equations for BCT to FCC are

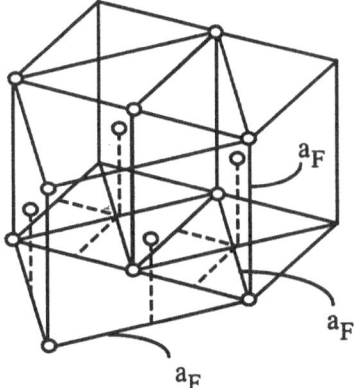

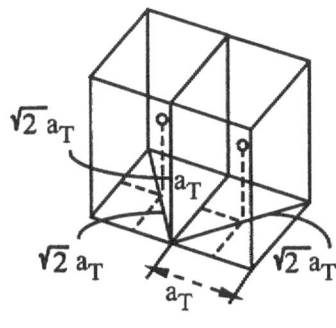

Figure 2.3. BCT basis for transformation to FCC.

Figure 2.4. FCC basis for transformation to BCT.

$$\mathbf{a_F} = a_T\,(\mathbf{i} + \mathbf{j}), \quad |\mathbf{a_F}| = \sqrt{2}\,a_T, \tag{2.14a}$$

$$\mathbf{b_F} = a_T\,(-\mathbf{i} + \mathbf{j}), \quad |\mathbf{b_F}| = \sqrt{2}\,a_T, \tag{2.14b}$$

$$\mathbf{c_F} = \mathbf{c_T}, \quad |\mathbf{c_F}| = \sqrt{2}\,a_T. \tag{2.14c}$$

The transformation equations for FCC to BCT are

$$\mathbf{a_T} = \frac{a_F}{2}\,(\mathbf{i} - \mathbf{j}), \quad |\mathbf{a_T}| = \frac{\sqrt{2}}{2}\,a_F, \tag{2.15a}$$

$$\mathbf{b_T} = \frac{a_F}{2}\,(\mathbf{i} + \mathbf{j}), \quad |\mathbf{b_T}| = \frac{\sqrt{2}}{2}\,a_F, \tag{2.15b}$$

$$\mathbf{c_T} = \mathbf{c_F}, \quad |\mathbf{c_T}| = a_F. \tag{2.15c}$$

The two cases are illustrated in Figures 2.3 and 2.4. For additional details on transformations from one system to another, see Chapter 6, where cases for primitive to centered transformations and some cases of special transformations are described in some detail. General transformations are discussed in [ITC, 1983].

2.2. Body-Centered Cubic

For BCC lattices, there are several definitions for the basis vectors. Usually, an orthogonal set is used which is nonprimitive, as shown in Figure 2.5. Buerger [1962] defined the primitive basis vector set as

$$\mathbf{a} = a\,\mathbf{i}, \qquad \mathbf{b} = a\,\mathbf{j}, \qquad \mathbf{c} = -\frac{a}{2}\mathbf{i} - \frac{a}{2}\mathbf{j} + \frac{a}{2}\mathbf{k}, \tag{2.16a}$$

$$\theta_{a,b} = 90°, \qquad \theta_{a,c} = \theta_{b,c} = 125.264...°, \tag{2.16b}$$

$$|\mathbf{a}| = |\mathbf{b}| = a, \qquad |\mathbf{c}| = \frac{a\sqrt{3}}{2}. \tag{2.16c}$$

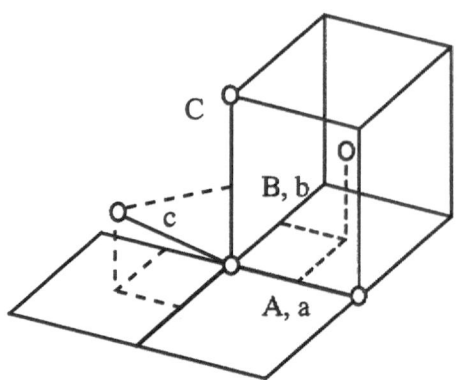

Figure 2.5. BCC defining primitive vectors **a**, **b**, **c** and body-centered vectors A, B, C.

A new set of vectors can be defined in terms of the primitive vectors as follows:

(a) $A = a,$ $B = b,$ $C = a + b + 2\,c,$ (2.17a)

$\theta_{A,B} = \theta_{B,C} = \theta_{C,A} = 90°,$ (2.17b)

$|A| = |B| = |C| = a.$ (2.17c)

Alternative defining sets from Ashcroft and Mermin [1976]:

(b) $a_1 = a\,\mathbf{i} = a,$ $a_2 = a\,\mathbf{j} = b,$ (2.18a)

$a_3 = \dfrac{a}{2}(\mathbf{i} + \mathbf{j} + \mathbf{k}) = a + b + c,$ (2.18b)

$|a_1| = |a_2| = a,$ $|a_3| = \dfrac{a\sqrt{3}}{2},$ (2.18c)

$\theta_{a1a2} = 90°,$ $\theta_{a2a3} = \theta_{a3a1} = 54.74°.$ (2.18d)

(c) $a_1 = \dfrac{a}{2}(-\mathbf{i} + \mathbf{j} + \mathbf{k}) = b + c,$ (2.19a)

$a_2 = \dfrac{a}{2}(\mathbf{i} - \mathbf{j} + \mathbf{k}) = a + c,$ (2.19b)

$a_3 = \dfrac{a}{2}(\mathbf{i} + \mathbf{j} - \mathbf{k}) = -c.$ (2.19c)

These basis vectors yield rhombohedral basis vectors with magnitudes and angles given by

$$|a_1| = |a_2| = |a_3| = \dfrac{a\sqrt{3}}{2},$$ (2.20a)

$$\theta_{a1a2} = \theta_{a2a3} = \theta_{a3a1} = 109.471...°,$$ (2.20b)

as shown in Figure 2.6.

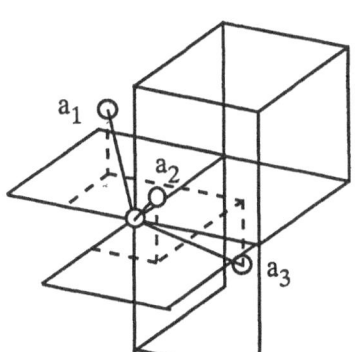

Figure 2.6. Rhombohedral basis vectors
derived from BCC basis vectors.

From Definition (a) and the **A, B, C** basis vector definitions above, the following sums show that for BCC lattices the sum of the indices must be even for the beams to appear:

$$H = h, \qquad K = k, \qquad L = h + k + 2\,l; \qquad\qquad (2.21a)$$

$$H + K + L = 2\,h + 2\,k + 2\,l = \text{even integer}, \qquad\qquad (2.21b)$$

where h, k, l are the indices associated with the primitive basis and H, K, L are the indices associated with the body centered basis.

For the rhombohedral bases using Definition (b), and H, K, L the rhombohedral indices, we have

$$
\begin{aligned}
H = h, \qquad K = k, \qquad L = h + k + l, \qquad &(2.22a)\\
H + K = h + k, \qquad &(2.22b)\\
K + L = h + 2\,k + l, \qquad &(2.22c)\\
L + H = 2\,h + k + l, \qquad &(2.22d)\\
H + K + L = 2\,h + 2\,k + l. \qquad &(2.22e)
\end{aligned}
$$

from which no general rules for allowed or extincted beams emerge for H, K, L. Special cases do yield rules: (h00): $L + H = 2n$; (0k0): $L + H = 2n$; (hk0): $H + K = 2n$.

For Definition (c), we have

$$
\begin{aligned}
H = k + l, \qquad K = h + l, \qquad L = -l, \qquad &(2.23a)\\
H + K = h + k + 2\,l, \qquad &(2.23b)\\
K + L = h, \qquad &(2.23c)\\
L + H = k, \qquad &(2.23d)\\
H + K + L = h + k + l, \qquad &(2.23e)
\end{aligned}
$$

which does not yield any general rules. Special rules are: (00l): $H + K = 2n$; $K = -L = H$; $H + K + L = l$.

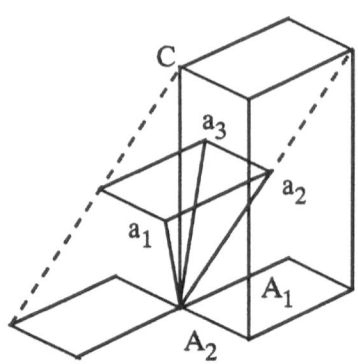

Figure 2.7. Hexagonal basis vector reference system with vectors A_1, A_2, C. Trigonal basis vectors derived from these are a_1, a_2, a_3 and angle α. [From M. J. Buerger, *X-Ray Crystallography*, John Wiley & Sons, New York, 1942, by permission.]

2.3. Hexagonal Close Packed

The hexagonal close packed (hcp) system poses special problems because of the use of several systems for indexing planes and directions. A full discussion of these indexing systems is presented in Chapter 5.

Defining vectors for hcp structures are used as a reference for trigonal and rhombohedral systems. Hence, in this section, two cases are considered: (1) rhombohedral or trigonal basis vectors, and (2) hexagonal basis vectors [Buerger, 1962].

Case (1): Rhombohedral basis vectors as reference: The reference basis vectors are a_1, a_2, a_3, which form a trigonal coordinate system, as shown in Figure 2.7.

Define the transformation from trigonal to hcp vectors as

$$A_1 = a_1 - a_2, \tag{2.24a}$$
$$A_2 = a_2 - a_3, \tag{2.24b}$$
$$C = a_1 + a_2 + a_3. \tag{2.24c}$$

To determine the allowed and extincted indices for this case, we note that if (hkl) are trigonal or rhombohedral plane indices, the basis vector transformation equations yield

$$H = h - k, \quad K = k - l, \quad L = h + k + l, \tag{2.25}$$

and

$$-H + K + L = 3k = \text{mod 3 integer.} \tag{2.26}$$

Hence, only plane indices satisfying this last equation will be present in reciprocal space.

Case 2: Hexagonal basis vector reference system: The reference vectors are A_1, A_2, C, which form a hexagonal coordinate system, as shown in Figure 2.7.

Define the transformation from hcp reference basis vectors A_1, A_2, C to trigonal vectors a_1, a_2, a_3, at angle α, as

$$a_1 = A_1 + C, \tag{2.27a}$$
$$a_2 = A_2 + C, \tag{2.27b}$$
$$a_3 = -A_1 - A_2 + C. \tag{2.27c}$$

Allowed and extincted beams are found by noting that

$$h = H + L, \quad k = K + L, \quad l = -H - K + L, \tag{2.28}$$

the sum of which is

$$h + k + l = \text{mod 3 integer.} \tag{2.29}$$

Hence, trigonal indices are mod 3 and will appear only when Equation (2.29) is satisfied.

2.4. End-Centered (Base-Centered)

The primitive basis vectors are a, b, c [Buerger, 1962], which in terms of the unit vector coordinate system are written as

$$a = \frac{A}{2}(i + j), \quad b = \frac{A}{2}(-i + j), \quad c = C\,k, \tag{2.30}$$

where A, B, C are as shown in Figure 2.8.

The transformation from the primitive basis vectors to the end-centered basis vectors is defined as

$$A = a - b, \quad B = a + b, \quad C = c, \tag{2.31}$$

where a, b, c are as shown in Figure 2.8.

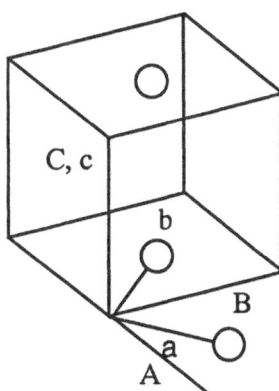

Figure 2.8. End-centered lattice primitive basis vectors a, b, c and nonprimitive vectors A, B, C.

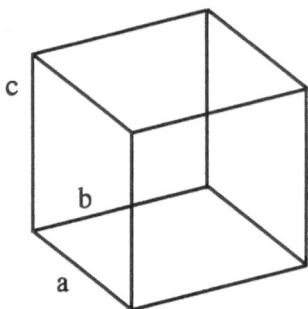

Figure 2.9. Simple cubic basis vectors
a, b, c.

From these equations, the plane indices transform as

$$H = h - k, \quad K = h + k, \quad L = l, \cdots \tag{2.32a}$$

$$H + K = 2h = \text{even integer.} \tag{2.32b}$$

Therefore, those plane indices with $H + K$ even appear, whereas those indices with $H + K$ odd do not.

2.5. Simple Cubic

The basis vectors for the primitive unit cell are given in terms of the unit reference orthogonal coordinate system

$$\mathbf{a} = a\mathbf{i}, \quad \mathbf{b} = b\mathbf{j}, \quad \mathbf{c} = c\mathbf{k}, \tag{2.33}$$

where **a, b, c** are as shown in Figure 2.9. No extinctions occur in this system.

3
Diffraction Pattern Analysis

3.1. Introduction

Diffraction pattern analysis consists of identifying the zone axis and the reciprocal lattice vectors in the pattern. This task requires knowledge of the properties of the crystal systems, the microscope, and some relationships among the measured distances, the measured angles, the reciprocal lattice, and the direct lattice.

Although there are many methods for accomplishing the analysis, each method includes measuring a minimum of three distances and the two included angles defined by these three distances. The camera constant must be known, although for some cases it is not necessary. The methods compare experimental distances and angles against theoretically calculated distances and angles using the equations presented in Chapter 5.

3.2. Errors in Measurements

3.2.1. Measurements on the emulsion

For hand measurements on diffraction patterns or convergent beam patterns and rings, the major sources of error are the ruler and protractor used. The assumption is that distortions due to the lenses in the microscope have been corrected before the plate was taken.

The measurements on film plates 60 x 90 mm in size is limited to about ±0.25 mm. Better accuracy can be obtained by measuring n beams on a line and dividing the measurement by n. The error is reduced to $m/\sqrt{n}$, where m is the distance measured on the film plate.

Angular measurements can be made to ± 0.5 degrees by using a well-made protractor. Mass-produced protractors may have a centering error, which is significant in terms of the error added to the hand measurement. Check any protractor used against a standard.

Photometer measurements can be made and the centers of beams located to better than 1% if the beams have not saturated the emulsion. If this approach is to be taken, then a series of standard exposures must be taken and used to determine the linearity of the film response. Once this is known, a series of exposures may be required to accurately measure beam separations.

Digitizers can be used to find beam positions on the plate. Sources of error here are limited to the users ability to locate the center of the beam and the visibility afforded by the mouse used. Some use a cross-hair marker, which distorts the beam position or obscures the beam completely. Reproducibility is about as good as hand measurement, but the advantage is that many measurements can be made and the statistics can be calculated using a computer.

3.2.2. Camera constant equation approximation

The camera constant equation is derived from the geometry of the optical axis and the image plane in relation to the wavevector axis and the reciprocal lattice vector. The assumption is that the wavevector and the optic axis are parallel.

For electron diffraction the Bragg angles, θ_B, given by

$$\frac{\lambda}{d} = 2 \sin \theta_B, \tag{3.1}$$

where λ is the electron wavelength and d is the interplanar spacing, are less than a few degrees, which allows the sine function to be approximated by the tangent function, giving,

$$\frac{\lambda}{d} \approx \tan (2 \theta_B) = \frac{R}{L}, \tag{3.2}$$

where R is the measured distance and L is the camera length. Expanding this in terms of the series representation gives

$$d = \frac{\lambda L}{R} \left\{ 1 + K \left(\frac{R}{L} \right)^2 + ... \right\}, \tag{3.3}$$

where K is 3/8 in the formal expansion, but is usually larger because of lens defect contributions [Reimer, 1988]. To a first approximation, this equation reduces to the camera constant equation relating the distance measured in the image, R, to the distance in the direct lattice, d:

$$R d = \lambda L. \tag{3.4}$$

3.2.3. Distortions in patterns

Distortions arise from misalignments in the intermediate and projector lens, spherical aberrations in the objective lens, and incorrect objective lens focusing.

Misalignments in the intermediate lens and projector lens can be corrected by adjusting the lens stigmators to produce round discs for slightly out-of-focus beams and by adjusting the condenser lens stigmator. Use medium camera lengths to be sure barrel distortion or other distortions that cause the image to be noncircular are minimized.

The specimen must be at the eucentric position and the objective aperture focused sharply. Operationally, these are the only two adjustments the user can make simply and quickly. The operator should, however, also adjust the image of the SAD (selected area diffraction) aperture to bring it into focus, which will also require refocusing the image. If the lens currents are available, these can be noted to allow reproducing conditions in another experiment.

The camera constant, λL, should be determined using a known standard such as Al or Au films. Once known, the camera constant will not change if the same conditions are reproduced for the specimen of interest. The camera constant for a given microscope should be known to an accuracy of at least 0.5%. Significant departures from this value indicate improperly set parameters in the microscope or that the batteries used in the electronics are bad or near exhaustion.

Reasonable care in measurements and calculations allow a precision of at least 2.5% and frequently 1%. Absolute values are more difficult to obtain, and uncertainties of 2% to 5% are not unusual if the absolute reference is data obtained by x-ray diffraction. A caution, however, must be stated: the thin foil may have relaxed lattice parameters, which introduce apparent errors of up to a few percent relative to the x ray values.

Spherical aberrations in the objective lens result in a displacement in the intermediate image plane of the image from the objective lens, i.e., the diffraction pattern. Williams [1983] gives the displacement as

$$Y_{sp} = C_s \, \alpha^3 \; + \text{small terms due to disc of confusion.} \qquad (3.5)$$

Hence the displacement is sensitive to aperture size, and beams at large projection angles may have sizable shifts.

For an improperly focused objective lens, the image plane is shifted away or back from the intended image plane, resulting in a shift of the position of the image source:

$$Y_f = D\alpha, \qquad (3.6)$$

where D = amount of shift of the focal plane.

The total shift is the sum of these two effects. The sign of D can be positive or negative (over or under focus); so by underfocusing, the total displacement can be minimized. For a Bragg angle of 0.015 rad, C_s = 1.6 mm, D = ± 2 μ, and λ = 0.0037 nm (100 kV), the displacement is [Thomas and Goringe, 1979, page 27]

$$Y_t = -25 \text{ nm or} + 35 \text{ nm.} \qquad (3.7)$$

3.2.4. Statistics and error

Error is the difference between the measured value and the true value [Beers, 1957]; the uncertainty in the measured value is represented by the standard deviation, or another measure of deviation. **Relative error** is the number produced by dividing the error by the average value. Multiplying by 100 gives the relative error in percent. **Precision** is the scatter about the measured value, regardless of the closeness of the measured value to the true value. **Accuracy** is a measure of the systematic error in a measurement. Small systematic error means high accuracy. **Systematic error** is a deviation from the true value exhibited in all the measurements made. The deviations are all in the same direction from the true value.

For n measurements of variable x with average value μ and standard deviation s, the error is

$$D = \frac{x - \mu}{s / \sqrt{n}} \; \frac{s}{\sqrt{n}} = z_{\alpha/2} \frac{s}{\sqrt{n}},$$ (3.8)

where $z_{\alpha/2}$ is the value of the normal distribution, found in tables of normal distributions [Miller and Freund, 1977].

When the number of measurements is less than 20, the Student's t distribution should be used. In this case the $z_{\alpha/2}$ is replaced by $t_{\alpha/2}$ the values for which are in the tables for Student t distribution.

The probability that the error is less than the estimate is defined as $1 - \alpha$, which is called the **confidence level** of the measurement.

If the measurements are of counts, as in x-ray or electron counts, the statistics follow the Poisson distribution, and the standard deviation [Goldstein and Yakowitz, 1975] is

$$\sigma_c = \left[\frac{\sum_{i=1}^{n} (N_i - \bar{N}_i)^2}{(n-1)} \right]^{1/2},$$ (3.9)

where N_i refers to the count in the ith collection of counts and $\bar{N}_i$ is the average count computed from

$$\bar{N}_i = \sum_{i=1}^{n} \frac{N_i}{n}.$$ (3.10)

The minimum standard deviation is the square root of the counts:

$$\sigma_c = \sqrt{N}.$$ (3.11)

3.3. Analysis of Patterns

3.3.1. Introduction

Identifying the planes associated with the spots present in a diffraction pattern is essential if quantitative analysis of the image features is to be done. In addition, to produce the optimum contrast, one must be able to obtain 2-beam conditions in a known way. Hence, even if semiquantitative information is required, knowledge of the diffraction conditions plays a major role in interpreting the features present in the image.

Determining the indices and the associated zone for a diffraction pattern is based on the convention that the +z axis is out of the paper toward the viewer; the axes are a right-handed system, and the emulsion side of the plate is up. In this convention,

Figure 3.1. Diagram showing the conventions used in analyzing and labeling a diffraction pattern for the +z convention.

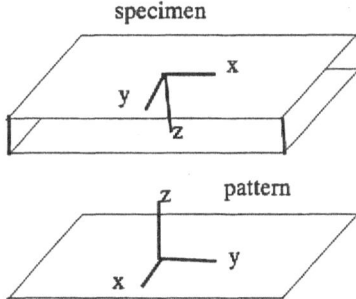

the zone can be found by taking the cross product of two reciprocal lattice vectors, as indicated in Figure 3.1.

Once two of the beams have been identified, the remaining beams can be indexed using vector addition, starting with the first two, and noting that the opposite beams have negative indices.

3.3.2. Indexing conventions

There is no universal indexing convention except for the requirement that the indices be consistent among themselves and that the axes used form a right-handed system.

3.3.3. Methods

There are two major approaches to identifying the planes present in a diffraction pattern. The first is applicable to ring patterns and the second to spot patterns.

3.3.3.1. Ring patterns

Ring patterns result from the presence of a large number of small grains in the area illuminated by the beam (see Figure 3.2). The rings actually consist of discrete beams from equivalent planes in each of the grains. The large population of grains

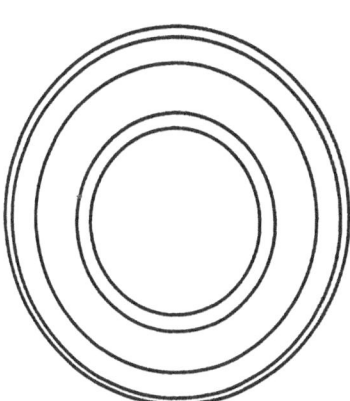

Figure 3.2. A ring pattern from an FCC structure. The rings are, from the inner to the outer ring, (111), (200), (220), (113), (222).

Figure 3.3. A plot of the square
root of the sum of the plane
indices versus R, the measured
distance on the negative plate.

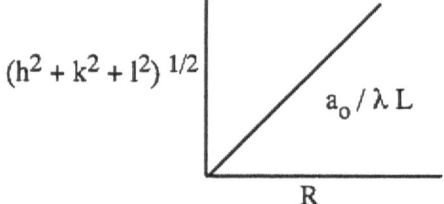

$(h^2 + k^2 + l^2)^{1/2}$

$a_o / \lambda L$

R

provides the statistics for producing a spot at a fixed distance $|\,g\,|$ from the
transmitted beam in a rotationally symmetric manner. The ring diameters, thus, are
proportional to $g(hkl)$, i.e., the diameters are proportional to the interplanar
separations present in the structures making up the grain.

The procedure for indexing the rings consists in measuring the diameters and
noting the expression relating $|\,g\,|$ to the indices and the lattice parameters of the
crystal system. For the cubic system we have

$$|\,g\,| = \frac{1}{d} = \frac{\sqrt{h^2 + k^2 + l^2}}{a_0}. \tag{3.12}$$

From the camera constant equation, write

$$\frac{R\,a_0}{L\lambda} = \sqrt{h^2 + k^2 + l^2}, \tag{3.13}$$

which is used by plotting values of $\sqrt{h^2 + k^2 + l^2}$ against values of the radius (Figure
3.3). The curve should be a straight line with slope $a_0 / L\,\lambda$. Because the camera
constant is known and R is measured, the lattice parameter can be found; if the lattice
parameter is known from other measurements such as x-ray diffraction, then the
camera constant can be determined.

Specific indexing of the rings is accomplished by taking ratios of radii and
comparing them against ratios of $\sqrt{h^2 + k^2 + l^2}$ for various combinations of the
indices. In ring patterns, the exact signs associated with the indices are not
significant.

3.3.3.2. Spot patterns

Although multiple patterns may be present, each spot diffraction pattern is produced
by a single crystal lattice. Hence, single patterns represent the structure of a domain
or a grain or a long period unit in a molecular chain. The pattern represents the
projection of the Ewald sphere onto the image plane. This reduction from three
dimensions (or higher) to a two-dimensional projection introduces ambiguity into the
interpretation of the pattern. Removal of this ambiguity can be done in several ways,
one of which is collection of many patterns from one grain. There is always,
however, a 180° ambiguity in any two-dimensional diffraction pattern that is
removable only by collecting patterns from other zones.

Indexing diffraction patterns can proceed in three ways: using ratios, using the zone and ratios, and using computer software. Obviously, the software approach is the most attractive because the work is minimized. However, a caution is required in using software. The final interpretation of any result, no matter how obtained, must reside with the experimenter, not with the software. There are many programs available for analyzing diffraction patterns, some of which run on PC or Macintosh type machines, others on larger machines. Before using any of them, be sure that the algorithm used is known and well understood, particularly its limitations.

Ratio method: The ratio method works exactly with cubic systems. Other systems can be solved, but the particulars of each structure limit the usefulness of the ratio method if the structure is not well known.

The method is similar to that employed for rings. Measure distances from the transmitted beam to the three shortest **g** vectors (diffracted beams), take the ratios of the measured distances and compare these against ratio tables for cubic systems. A consistent set usually can be found if three distinct ratios are used.

As an example, consider the pattern obtained from an Al specimen. Measurement of distances marked A, B, C in this pattern (Figure 3.4) give values

$$A = 57.6/6 = 9.6 \text{ mm,} \qquad (3.14a)$$
$$B = 48/2 \;\; = 24.0 \text{ mm,} \qquad (3.14b)$$
$$C = 49/2 \;\; = 24.5 \text{ mm.} \qquad (3.14c)$$

From these, obtain the ratios

$$B/A = 2.50, \qquad (3.15a)$$
$$C/B = 1.02, \qquad (3.15b)$$
$$C/A = 2.55. \qquad (3.15c)$$

Examination of the table for FCC ratios shows the following set to satisfy the measured ratios:
$$331/111 = 2.52, \qquad (3.16a)$$
$$420/331 = 1.03, \qquad (3.16b)$$
$$420/111 = 2.58. \qquad (3.16c)$$

Hence the planes are for A: {111}; for B: {331}; for C: {420} .

The next task, now that the plane types have been found, is to obtain a set of indices that is internally consistent, that is, the indices must be obtainable by vector addition of the **g** vectors in the pattern. Since the crystal does not really care what we call the planes, we must decide the minimum number of specific planes and

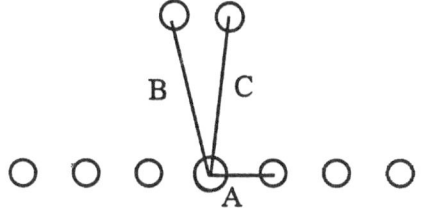

Figure 3.4. Diffraction pattern from Al.

derive the remainder from these. The minimum number needed is two, since by addition a third can be obtained, and, from this, a fourth, and so on till all planes are identified. In this example, choose the {111} and the {331} families to start with. Choose $(1\bar{1}1)$ and $(33\bar{1})$ as the starting specific indices. Adding these gives (420), the third plane identified in the analysis.   Other planes can be found quickly by noting that planes symmetrically opposite through the transmitted beam have complimentary indices. Thus the plane opposite to $(\bar{1}\bar{1}1)$ is $(\bar{1}1\bar{1})$, to $(33\bar{1})$ is $(\bar{3}\bar{3}1)$, and to (420) is $(\bar{4}\bar{2}0)$.

Zone Axis Method: The zone axis method is the more common and general approach. The pattern in the previous example is solved by this method by noting not only the distances but also the two included angles, $AC = 75$ and $BC = 22.6$. Using the camera constant, convert the measured distances to d spacings. Use these and a list of d spacings for Al to identify the plane types . Pick two of the planes and find the third by vector addition. Calculate the angle between the pairs of planes using the specific indices chosen. Compare these with the measured angles. If they agree, the planes have been consistently identified. The zone is calculated using the cross product of two of the planes, the specific direction indices being the reduced set

of [uvw]. In the example the zone is $[\bar{1}23]$. **The analysis is not complete until agreement with the distances and the angles is confirmed and the zone calculated.**

Summary: (1) Identification of a pattern is not necessarily unique. Different crystal systems can yield similar patterns, and different centerings can produce identical patterns, e.g., BCC and FCC can be misinterpreted unless one knows the centering of the phase before analyzing the pattern. (2) If the phase is known, the ambiguity of (1) is removed or reduced. (3) If the phase is not known, obtain data on possible systems from compilations of crystallographic data, literature references, or any place where the reliability of the data is verifiable.

3.4. Structure Factor

Given a lattice of points with origins at r_0 and atoms within the unit cell at r_t. The Fourier transform of the sum over the points is

$$F(r)\, G(r)= \sum_{t=1}^{n} f_t \exp\left[-\,2\,\pi\, i\,(k-k_0) \cdot r_0\right]$$
$$\times \sum_{u}^{M_u} \sum_{v}^{M_v} \sum_{w}^{M_w} \exp\left[-\,2\,\pi\, i\,(k-k_0) \cdot r_0\right]. \qquad (3.17)$$

The first factor represents the structure factor, dependent on the positions of the atoms in the unit cell. The scattering by each atom is represented by f_t, the atom scattering amplitude, values of which are tabulated in the International Crystallographic Tables [ITXC, 1959]. The wavevector after and before scattering is represented by k and k_0. The second factor is the shape factor and depends on the number of cells scattering the beam. The number of cell repeats along the basis axes are M_u, M_v, M_w.

The structure factor, F, has significant value only near lattice points. The shape factor is nonzero near lattice points, the width being proportional to the reciprocal of the number of scattering unit cells.

This formulation is strictly suitable only for kinematic conditions (single scattering, very thin specimen).

Table 3.1. Some Values of Structure Factors $|F|^2$.

$|F|^2$	Condition	$F|^2$	Condition
FCC (A)		Sodium	
0	h, k, l mixed parity	Chloride (AB)	
$16 f_A^2$	h, k, l all even or odd	0	h, k, l mixed parity
BCC (A)		$16 (f_A - f_B)^2$	h, k, l all odd
0	h + k + l = odd	$16 (f_A + f_B)^2$	h, k, l all even
$4 f_A^2$	h + k + l = even		
		Zincblende (AB)	
Diamond		0	h, k, l mixed parity
Cubic (A)		$16 (f_A^2 + f_B^2)$	h, k, l all odd
0	h, k, l mixed parity	$16 (f_A + f_B)^2$	h, k, l all even, h + k + l = 4n
0	h, k, l all even, h + k + l = 4n +2		
$32 f_A^2$	h, k, l all odd	$16 (f_A - f_B)^2$	h, k, l all even, h + k + l = 4n +2
$64 f_A^2$	h, k, l all even, h + k + l = 4n		
		hcp (A)	
Cesium		0	l odd, h + 2 k = 3n
Chloride (AB)		f_A^2	l even, h + 2 k = 3n + 1 or 3n + 2
$(f_A + f_B)^2$	h + k +l = even		
$(f_A - f_B)^2$	h + k + l = odd	$3 f_A^2$	l odd, h + 2 k = 3n +1 or 3n + 2
		$4 f_A^2$	l even, h + 2 k = 3n

After Reimer [1988].

The structure factor takes on certain characteristic forms for the simple lattices. Examples are listed in Table 3.1 [Reimer, 1988]. The exact forms for all the space groups are detailed in the 1952 edition of the ITXC.

3.5. Ratio Tables

The tables of plane ratios for FCC, BCC, SC, and hcp for l = 0 are given in Tables 3.2, 3.3, 3.4, 3.5. These tables represent universal ratios applicable to any material possessing the appropriate crystal system. The remaining crystal systems do not exhibit universal tables except by accident because the ratios of basis repeat distances differ among the three axes. Planes in the tetragonal system with l = 0, i.e., (hk0) planes do exhibit a universal behavior. These are listed in Table 3.6. The notation used is $(1/d_2)/(1/d_1)$ as a reminder that the measured distances on the negative are reciprocal values. Equivalent notation is g_2/g_1.

For the hcp system, the equation relating the c/a ratio (γ) to any pair of planes is given by

$$\gamma^2 = \frac{3\,(l_1 - l_2\,d_{21}{}^2)}{4\,[(h_2{}^2 + k_2{}^2 + h_2\,k_2)\,d_{21}{}^2 - (h_1{}^2 + k_1{}^2 + h_1\,k_1)]}, \tag{3.18}$$

where ($h_1\ k_1\ l_1$)and ($h_2\ k_2\ l_2$) are indicess of any two planes, and

$$d_{21} = \frac{(1/d_2)}{(1/d_1)}, \tag{3.19}$$

is the ratio of experimentally measured reciprocal lattice vector distances in the diffraction pattern. Using this equation, the experimental value of c/a can be checked for consistency among a number of measured values from the diffraction pattern.

The corresponding equation for the tetragonal system is

$$\gamma^2 = \frac{3\,(l_1 - l_2\,d_{21}{}^2)}{4\,[(h_2{}^2 + k_2{}^2)\,d_{21}{}^2 - (h_1{}^2 + k_1{}^2)]}. \tag{3.20}$$

Table 3.2. Ratio Tables for FACE-CENTERED CUBIC Lattice.

$(1/d_2) / (1/d_1)$

$(1/d_2)$	111	200	220	311	222	400	331	420	422	333
$(1/d_1)$:	1.732	2.000	2.828	3.317	3.464	4.000	4.359	4.472	4.899	5.196
111 1.732	1.000	1.155	1.633	1.915	2.000	2.309	2.517	2.582	2.828	3.000
200 2.000	0.866	1.000	1.414	1.658	1.732	2.000	2.179	2.236	2.449	2.598
220 2.828	0.612	0.707	1.000	1.173	1.225	1.414	1.541	1.581	1.732	1.837
311 3.317	0.522	0.603	0.853	1.000	1.044	1.206	1.314	1.348	1.477	1.567
222 3.464	0.500	0.577	0.816	0.957	1.000	1.155	1.258	1.291	1.414	1.500
400 4.000	0.433	0.500	0.707	0.829	0.866	1.000	1.090	1.118	1.225	1.299
331 4.359	0.397	0.459	0.649	0.761	0.795	0.918	1.000	1.026	1.124	1.192
420 4.472	0.387	0.447	0.632	0.742	0.775	0.894	0.975	1.000	1.095	1.162
422 4.899	0.354	0.408	0.577	0.677	0.707	0.816	0.890	0.913	1.000	1.061
333* 5.196	0.333	0.385	0.544	0.638	0.667	0.770	0.839	0.861	0.943	1.000

*333 = 511 also

Table 3.3. Ratio Tables for BODY-CENTERED CUBIC Lattice.

$(1/d_2)/(1/d_1)$

$(1/d_2)$	110	200	211	220	310	222	321	400	411	420	332	422
$(1/d_1)$:	1.414	2.000	2.449	2.828	3.162	3.464	3.742	4.000	4.243	4.472	4.690	4.899
110 1.414	1.000	1.414	1.732	2.000	2.236	2.449	2.646	2.828	3.000	3.162	3.317	3.464
200 2.000	0.707	1.000	1.225	1.414	1.581	1.732	1.871	2.000	2.121	2.236	2.345	2.449
211 2.449	0.577	0.816	1.000	1.155	1.291	1.414	1.528	1.633	1.732	1.826	1.915	2.000
220 2.828	0.500	0.707	0.866	1.000	1.118	1.225	1.323	1.414	1.500	1.581	1.658	1.732
310 3.162	0.447	0.632	0.775	0.894	1.000	1.095	1.183	1.265	1.342	1.414	1.483	1.549
222 3.464	0.408	0.577	0.707	0.816	0.913	1.000	1.080	1.155	1.225	1.291	1.354	1.414
321 3.742	0.378	0.535	0.655	0.756	0.845	0.926	1.000	1.069	1.134	1.195	1.254	1.309
400 4.000	0.354	0.500	0.612	0.707	0.791	0.866	0.935	1.000	1.061	1.118	1.173	1.225
411* 4.243	0.333	0.471	0.577	0.667	0.745	0.816	0.882	0.943	1.000	1.054	1.106	1.155
420 4.472	0.316	0.447	0.548	0.632	0.707	0.775	0.837	0.894	0.949	1.000	1.049	1.095
332 4.690	0.302	0.426	0.522	0.603	0.674	0.739	0.798	0.853	0.905	0.953	1.000	1.044
422 4.899	0.289	0.408	0.500	0.577	0.645	0.707	0.764	0.816	0.866	0.913	0.957	1.000

* 411= 330 also

Table 3.4. Ratio Tables for Simple Cubic Lattice.
$(1/d_2)/(1/d_1)$

(1/d₂)		100	110	111	200	210	211	220	221	310	311	222	320	321	400
(1/d₁):		1.000	1.414	1.732	2.000	2.236	2.449	2.828	3.000	3.162	3.317	3.464	3.606	3.742	4.000
100	1.000	1.000	1.414	1.732	2.000	2.236	2.449	2.828	3.000	3.162	3.317	3.464	3.606	3.742	4.000
110	1.414	0.707	1.000	1.225	1.414	1.581	1.732	2.000	2.121	2.236	2.345	2.449	2.550	2.646	2.828
111	1.732	0.577	0.816	1.000	1.155	1.291	1.414	1.633	1.732	1.826	1.915	2.000	2.082	2.160	2.309
200	2.000	0.500	0.707	0.866	1.000	1.118	1.225	1.414	1.500	1.581	1.658	1.732	1.803	1.871	2.000
210	2.236	0.447	0.632	0.775	0.894	1.000	1.095	1.265	1.342	1.414	1.483	1.549	1.612	1.673	1.789
211	2.449	0.408	0.577	0.707	0.816	0.913	1.000	1.155	1.225	1.291	1.354	1.414	1.472	1.528	1.633
220	2.828	0.354	0.500	0.612	0.707	0.791	0.866	1.000	1.061	1.118	1.173	1.225	1.275	1.323	1.414
221*	3.000	0.333	0.471	0.577	0.667	0.745	0.816	0.943	1.000	1.054	1.106	1.155	1.202	1.247	1.333
310	3.162	0.316	0.447	0.548	0.632	0.707	0.775	0.894	0.949	1.000	1.049	1.095	1.140	1.183	1.265
311	3.317	0.302	0.426	0.522	0.603	0.674	0.739	0.853	0.905	0.953	1.000	1.044	1.087	1.128	1.206
222	3.464	0.289	0.408	0.500	0.577	0.645	0.707	0.816	0.866	0.913	0.957	1.000	1.041	1.080	1.155
320	3.606	0.277	0.392	0.480	0.555	0.620	0.679	0.784	0.832	0.877	0.920	0.961	1.000	1.038	1.109
321	3.742	0.267	0.378	0.463	0.535	0.598	0.655	0.756	0.802	0.845	0.886	0.926	0.964	1.000	1.069
400	4.000	0.250	0.354	0.433	0.500	0.559	0.612	0.707	0.750	0.791	0.829	0.866	0.901	0.935	1.000

*221 = 300 also

Table 3.5. Plane Ratios for the Tetragonal System for (hk0) Planes.
$(1/d_2)/(1/d_1)$

(1/d₂)				100	200	300	400	110	120	130	140	220	230	240	340
(1/d₁)				1.000	2.000	3.000	4.000	1.414	2.236	3.162	4.123	2.828	3.606	4.472	5.000
1	0	0	1.000	1.000	2.000	3.000	4.000	1.414	2.236	3.162	4.123	2.828	3.606	4.472	5.000
2	0	0	2.000	0.500	1.000	1.500	2.000	0.707	1.118	1.581	2.062	1.414	1.803	2.236	2.500
3	0	0	3.000	0.333	0.667	1.000	1.333	0.471	0.745	1.054	1.374	0.943	1.202	1.491	1.667
4	0	0	4.000	0.250	0.500	0.750	1.000	0.354	0.559	0.791	1.031	0.707	0.901	1.118	1.250
1	1	0	1.414	0.707	1.414	2.121	2.828	1.000	1.581	2.236	2.915	2.000	2.550	3.162	3.536
1	2	0	2.236	0.447	0.894	1.342	1.789	0.632	1.000	1.414	1.844	1.265	1.612	2.000	2.236
1	3	0	3.162	0.316	0.632	0.949	1.265	0.447	0.707	1.000	1.304	0.894	1.140	1.414	1.581
1	4	0	4.123	0.243	0.485	0.728	0.970	0.343	0.542	0.767	1.000	0.686	0.874	1.085	1.213
2	2	0	2.828	0.354	0.707	1.061	1.414	0.500	0.791	1.118	1.458	1.000	1.275	1.581	1.768
2	3	0	3.606	0.277	0.555	0.832	1.109	0.392	0.620	0.877	1.144	0.784	1.000	1.240	1.387
2	4	0	4.472	0.224	0.447	0.671	0.894	0.316	0.500	0.707	0.922	0.632	0.806	1.000	1.118
3	4	0	5.000	0.200	0.400	0.600	0.800	0.283	0.447	0.632	0.825	0.566	0.721	0.894	1.000

Table 3.6. Ratio Tables for Hexagonal Close Packed Lattice for l = 0.
$(1/d_2)/(1/d_1)$

(1/d₁)				(1/d₂)	01-10	11-20	12-30	13-40	20-20	22-40	23-50	24-60
h	k	i	l		1.155	2.000	3.055	4.163	2.309	4.000	5.033	6.110
0	1	-1	0	1.155	1.000	1.732	2.646	3.606	2.000	3.464	4.359	5.292
1	1	-2	0	2.000	0.577	1.000	1.528	2.082	1.155	2.000	2.517	3.055
1	2	-3	0	3.055	0.378	0.655	1.000	1.363	0.756	1.309	1.648	2.000
1	3	-4	0	4.163	0.277	0.480	0.734	1.000	0.555	0.961	1.209	1.468
2	0	-2	0	2.309	0.500	0.866	1.323	1.803	1.000	1.732	2.179	2.646
2	2	-4	0	4.000	0.289	0.500	0.764	1.041	0.577	1.000	1.258	1.528
2	3	-5	0	5.033	0.229	0.397	0.607	0.827	0.459	0.795	1.000	1.214
2	4	-6	0	6.110	0.189	0.327	0.500	0.681	0.378	0.655	0.824	1.000

4
Packing Fraction and Stacking Sequences

4.1. Packing Fraction

The packing of atoms to form a structure can be accomplished in a number of ways. With respect to the 14 Bravais lattices, the packing (and the stacking of layers) is relatively simple. There are "privileged" positions in cubic and hcp stacking, designated as A, B, or C sites, planes, or stacking. The geometrical definitions of A, B, and C are given in Figure 4.1.

The definition of packing fraction is

$$\text{Packing fraction} = \frac{\text{(volume of number of atoms in a unit cell)}}{\text{volume of unit cell}}. \tag{4.1}$$

4.1.1. Simple cubic

Assume the lattice is composed of spheres with radius = 1/2 unit. Then the volume is

$$V_{\text{sphere}} = \frac{\pi}{6}. \tag{4.2}$$

For the simple cubic lattice, close packing means the atoms touch along the edges of the cube as suggested in Figure 4.2(a). Since the edges are each 1 unit in length, the volume is $V_{\text{cube}} = 1$. The number of spheres per unit cell is obtained by counting the number of spheres contained within the boundaries of the unit cell. For the simple cubic unit cell, there are 8 spheres, one located at each corner. The fraction of each sphere located within the cube boundaries is 1/8. Hence, there is 1 sphere per unit cell ($8 \cdot 1/8 = 1$), and the packing fraction is

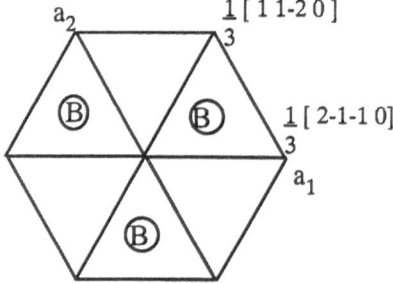

Figure 4.1. Definition of the ABC positions in the (111) projection of a cubic cell or a (0001) projection of an hcp cell. The A position lies at the center, the B positions, as shown, and the C positions in the centers of the empty triangles. a_1 and a_2 refer to the basis vectors for the (3, 3) or (4, 4) hcp axes.

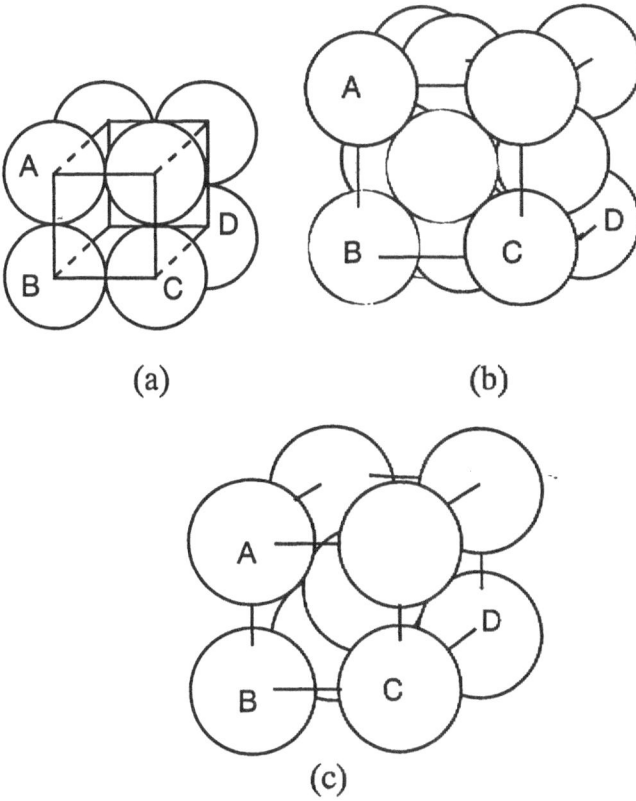

Figure 4.2. Unit cells with spheres of radius 1/2 unit length. (a) Simple cubic: spheres touch along the edges; the distances along the edges are each 1 unit in length; (b) face-centered cubic: the spheres touch on the face diagonals; the edge lengths are each $\sqrt{2}$ units; (c) body-centered cubic: the spheres touch along the cube diagonals; the edge lengths are each $2/\sqrt{3}$ units in length.

$$\{\text{packing fraction}\} = \left\{ \frac{\text{no. of spheres}}{\text{unit cell}} \right\} \frac{\{\text{vol. of 1 sphere}\}}{\text{vol. of unit cell}}$$

$$= (1)\left(\frac{\pi}{6}\right)\frac{(1)}{(1)} = \frac{\pi}{6}. \tag{4.3}$$

4.1.2. Face-centered cubic

For FCC, close packing means the spheres touch on the face diagonals [see Figure 4.2(b)]. Since three spheres touch on a diagonal, the length of the diagonal is 2 units, and, therefore, the cube edges are $\sqrt{2}$ units. The volume of the cube is $2\sqrt{2}$ units^3. There are 4 spheres per unit cell, i.e., 1/8 of a sphere at each corner times 8 corners plus 1/2 sphere times six faces. Therefore,

Table 4.1. Unit Cell Parameters for SC, FCC, and BCC (after Kittel [1967, p. 18, table 3]).

Parameters	SC	FCC	BCC
Lattice parameters	a_{SC}	$a_{FCC} = \sqrt{2}\, a_{SC}$	$a_{BCC} = \dfrac{2\, a_{SC}}{\sqrt{3}}$
Volume of conventional cell	a_{SC}^3	a_{FCC}^3	a_{BCC}^3
Volume of the primitive cell	a_{SC}^3	$\dfrac{a_{FCC}^3}{4}$	$\dfrac{a_{BCC}^3}{2}$
Lattice points/cell	1	4	2
Number of nearest neighbors	6	12	8
Nearest neighbor distance	a_{SC}	$\dfrac{\sqrt{2}\, a_{FCC}}{2} = a_{SC}$	$\dfrac{\sqrt{3}\, a_{BCC}}{2} = a_{SC}$
Number of second nearest neighbors	12	6	6
Second nearest neighbor distance	$\sqrt{2}\, a_{SC}$	$a_{FCC} = \sqrt{2}\, a_{SC}$	$a_{BCC} = \dfrac{2\, a_{SC}}{\sqrt{3}}$
Packing fraction	$\dfrac{\pi}{6}$	$\sqrt{2}\,\dfrac{\pi}{6}$	$\sqrt{3}\,\dfrac{\pi}{6}$

$$\{\text{packing fraction}\} = (4)\left(\frac{\pi}{6}\right)\frac{(1)}{(2\sqrt{2})} = \sqrt{2}\,\frac{\pi}{6}, \quad \text{FCC.} \tag{4.4}$$

4.1.3. Body-centered cubic

For BCC close packing means the spheres touch along the cube body diagonal [Figure 4.2(c)]. Hence, the diagonal is 2 units, and the sides are $2/\sqrt{3}$ units. There are 2 spheres per unit cell, i.e., 1/8 sphere at the 8 corners and 1 in the center. Therefore,

$$\{\text{packing fraction}\} = (2)\left(\frac{\pi}{6}\right)\frac{(1)}{\left(\frac{8}{3\sqrt{3}}\right)} = \sqrt{3}\,\frac{\pi}{8}, \quad \text{BCC.} \tag{4.5}$$

4.1.4. Hexagonal close packed

For hexagonal close packing, the atoms touch such that the centers of three coplanar atoms form a triangle of side twice the radius of the spheres. The angles of the triangle are all equal and are equal to $60°$. For close packing in three dimensions, the second layer must lie so that the distance between atoms is minimized. As shown in Figure 4.3, the geometry is such that

$$\frac{c^2}{4} + \frac{a^2}{3} = a^2. \tag{4.6}$$

Thus, ideal packing implies

$$\frac{c}{a} = \sqrt{\frac{8}{3}}, \tag{4.7}$$

where a = diameter of spheres, c = separation of *equivalent* planes of spheres.

4.2. Stacking Sequences

4.2.1. Simple cubic

The close packed plane is (100), and the stacking is ...XXXXX...; adjacent plane separation is a. See Figure 4.4.
    Along [110], the stacking is ...XYXYXY...; adjacent plane separation is a$\sqrt{2}$/2. See Figure 4.5.
    Along [111], the stacking is ...XYZXYZXYZ...; adjacent plane separation is a$\sqrt{3}$/3. See Figure 4.6.
    The two-dimensional projection onto the origin plane of the plane stacking can be obtained by determining the projected coordinates of an atom from the plane to be projected. Once this shift vector is known, the entire plane is shifted as a rigid

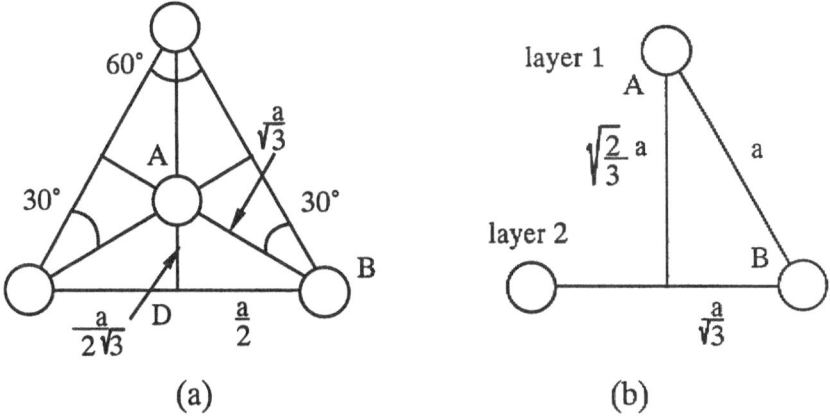

(a)   (b)

Figure 4.3. (a) Geometry of the three centers of atoms or lattice points in a close packed arrangement; viewed along the cube diagonal of a simple cube; (b) geometry perpendicular to (a) along the line AB showing the close packing between adjacent planes. The circles represent the projection of lattice points. If the stacking distance differs from a$\sqrt{2/3}$ and if the B points are located in the C positions shown in Figure 4.1,then the layers are stacked as in the hcp stacking and the separation between layers becomes c/2, where c is the hcp lattice parameter associated with the [0001] direction.

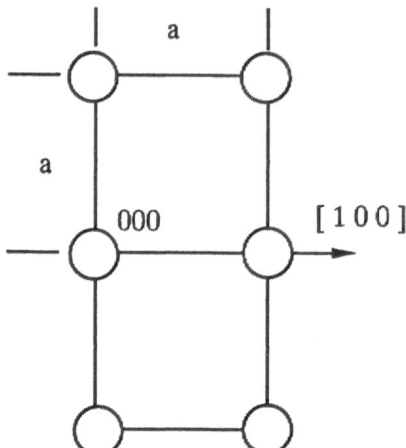

Figure 4.4. Simple cubic plane projection along [001]; the lattice sites in each plane along [100] shadow the sites below them.

construction. The result is the projection of that plane onto the origin plane. See Chapter 8 for details. The shift vector for a [111] projection is determined by noting that the atom coordinates of the [100] position in the first plane (Y) above the origin and parallel to it are shifted by a vector of type $\frac{1}{3}[2\bar{1}\bar{1}]$.

Atom coordinates of [110] position for the second plane (Z) above the origin are translated by vectors of type $\frac{1}{3}[11\bar{2}]$.

4.2.2. Body-centered cubic

The close packed plane is (110), and the stacking is ...XYXYX... . The adjacent plane separation is $a\sqrt{2}/2$.

Along [100], stacking is ...XYXYXYXY...; the adjacent plane separation is a/2.

Along [111], stacking is ...XYZUVW...; the adjacent plane separation is $a\sqrt{3}/6$.

Along [112], stacking is ...XYZUVW...; the adjacent plane separation is $a\sqrt{6}/6$.

Note that, in determining the stacking in the [111] direction, if the planes between [000] and [111] are counted, the sequence is ...XYZUVW..., whereas if the planes between [000] and the body center plane at $[\frac{1}{2}\frac{1}{2}\frac{1}{2}]$ are counted, the sequence is ...ABC..., which is identical to FCC. Hence the FCC sequence appears every other set of three planes.

4.2.3. Face-centered cubic

Close-packed planes are {111} along <111>, and the stacking is ...ABCABCA...; the adjacent plane separation is $a\sqrt{3}/3$.

Along <100>, the stacking is ...ABABAB...; the adjacent plane separation is a/2.

Along <110>, the stacking is ...ABABABA...; the adjacent plane separation is $a\sqrt{2}/4$.

Since some confusion may arise because of the common use of A, B, C for both planes and specific sites, the same stacking is written as numbers ...123..., where each number represents a plane in the adjacent plane sequence, which it must be noted is not necessarily the equivalent plane sequence.

Thus, in the numerical notation, close packed planes are {111} along <111>, and the stacking is ...1231231... .

Along <100>, the stacking is ...121212... .

Along <110>, the stacking is ...1212121... .

4.2.4. Hexagonal close packed

Close-packed planes are {0001} along <0001>, and the stacking is ...ABABA... . Adjacent plane spacing is c/2.

4.3. Interstitial Positions

The octahedral and tetrahedral sites in the FCC, BCC, and hcp lattices play important roles in bonding, strength, and other properties of materials. Details of the exact positions in these three lattices were extracted from Barrett and Massalski [1980]. Other sources of value are Pearson [1972] and Parthe [1964].

4.3.1. Face-centered cubic

Octahedral sites have 6 atoms/molecules in coordination. In the FCC unit cell there are 13 octahedral possibilities: 12 are at midpoints of cell edges, and 1 lies at the center of the unit cell, but note that there are 4 complete sites in a unit cell.

Tetrahedral sites have 4 atoms or molecules in coordination. In the FCC unit cell there are 8 tetrahedral sites: 2 are located on each body diagonal (a total of 4 x 2 = 8) at coordinates $\frac{1}{4},\frac{1}{4},\frac{1}{4}$ and $\frac{3}{4},\frac{3}{4},\frac{3}{4}$. See Figure 4.7.

4.3.2. Body-centered cubic

Octahedral sites have six atoms/molecules in coordination. In the BCC unit cell there are 18 octahedral sites: 6 are located in face centers, and 12 are located on cell edges, but note that there are 4 complete sites in a unit cell.

Tetrahedral sites have four atoms in coordination: In the BCC unit cell there are 24 sites (4 sites on each face x 6 faces = 24 sites). See Figure 4.8.

4.3.3. Hexagonal close packed

Octahedral sites have six atoms or molecules in coordination. In the hcp unit cell there are 6 octahedral sites. Three sites are at 1/3 z and three sites are at 2/3 z.

Tetrahedral sites have four atoms or molecules in coordination. In the hcp unit cell there are 20 tetrahedral sites: 14 sites lie on the hexagonal cell edges, 2 lie on central axes, 3 lie on lower sites, and 3 lie on upper sites.

The interstitial sites are shown in Figures 4.9 and 4.10 for the ideal hcp ratio

$\sqrt{8/3}$ = 1.632993.

Geometric details of the octahedra and tetrahedra for FCC, BCC, and hcp interstitial positions are given in Figure 4.11.

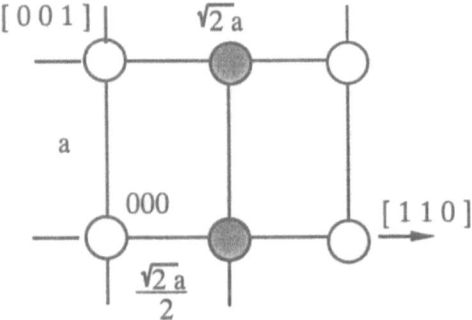

Figure 4.5. Simple cubic plane projection along [1̄10]; the lattice sites in plane 1 shadow the lattice sites in plane 3; sites in plane two, indicated by the cross-hatched circles, are shifted, lying $a\sqrt{2}/2$ below and above the page, and are frequently designated as B sites to indicate that adjacent planes are shifted relative to the reference plane, designated as plane A and which contains A sites.

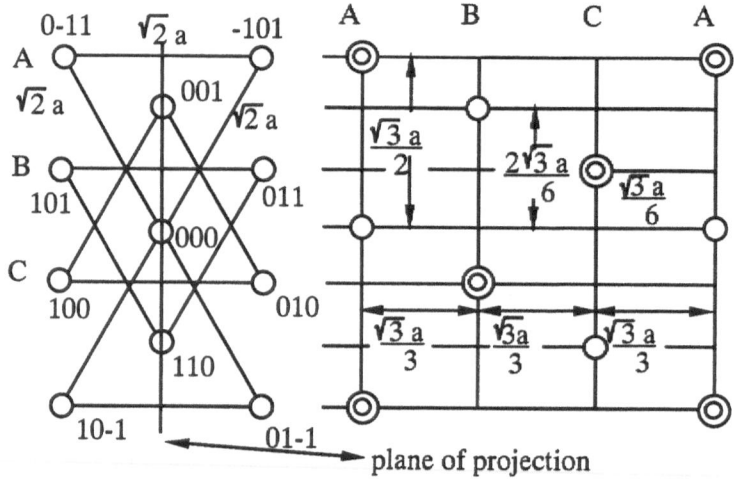

Figure 4.6. (a) Simple cubic projection along the [111] direction. (b) Projection along [1 1̄0] showing the relative location of lattice sites in planes relative to the reference plane A. In the [111] direction, plane A sites shadow sites in the third plane away. Hence, there are two planes between equivalent planes A, and each of these planes is shifted with respect to the reference A plane. The sites in these planes are designated as B sites for plane 2 and C sites for plane 3. These planes are also designated as B planes and C planes.

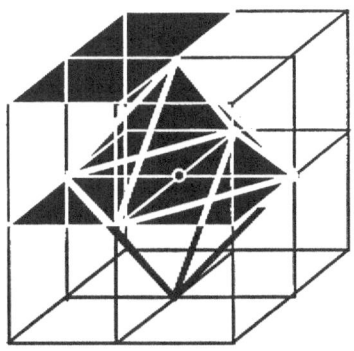

Figure 4.7. Interstitial sites in the FCC structure: (a) Tetrahedral sites on body diagonals; second sites are located at symmetric positions on the same diagonal; there are a total of 8 sites (2 per diagonal x 4 diagonals). (b) Octahedral sites on the edges of the cube; there are a total of 13 octahedral sites.

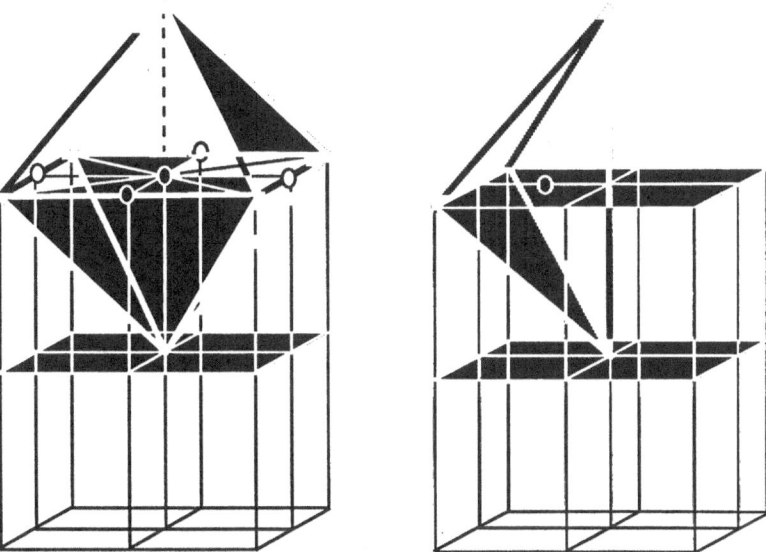

Figure 4.8. Interstitial sites in the BCC structure: (a) Octahedral sites on one face; there are 18 sites (2x6 edges + 6 faces). (b) Tetrahedral sites on each face; there are 24 sites (4 per face x 6 faces).

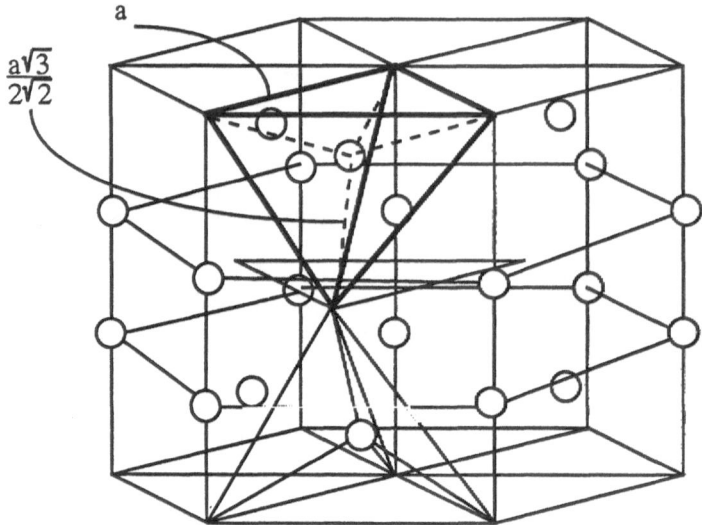

Figure 4.9. Interstitial sites in hcp: tetrahedral sites in the hexagonal close packed unit cell. There are a total of 20 sites.

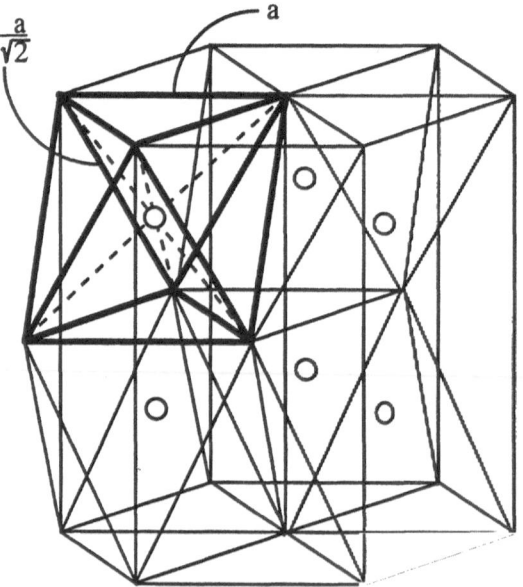

Figure 4.10. Interstitial sites in the hcp structure: octahedral sites in the hexagonal close packed unit cell. There are a total of six sites.

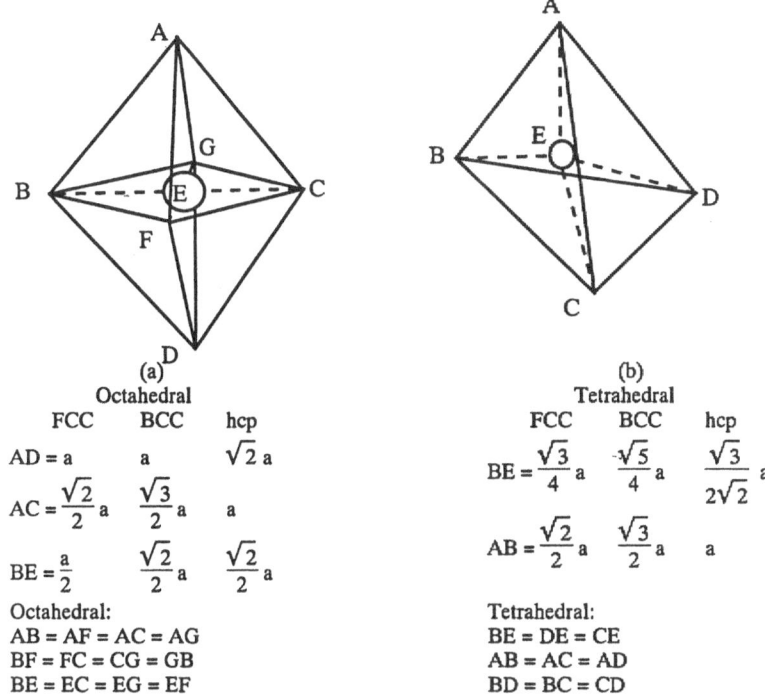

(a)
Octahedral

FCC	BCC	hcp
$AD = a$	a	$\sqrt{2}\,a$
$AC = \dfrac{\sqrt{2}}{2}\,a$	$\dfrac{\sqrt{3}}{2}\,a$	a
$BE = \dfrac{a}{2}$	$\dfrac{\sqrt{2}}{2}\,a$	$\dfrac{\sqrt{2}}{2}\,a$

Octahedral:
AB = AF = AC = AG
BF = FC = CG = GB
BE = EC = EG = EF

(b)
Tetrahedral

FCC	BCC	hcp
$BE = \dfrac{\sqrt{3}}{4}\,a$	$\dfrac{\sqrt{5}}{4}\,a$	$\dfrac{\sqrt{3}}{2\sqrt{2}}\,a$
$AB = \dfrac{\sqrt{2}}{2}\,a$	$\dfrac{\sqrt{3}}{2}\,a$	a

Tetrahedral:
BE = DE = CE
AB = AC = AD
BD = BC = CD

Figure 4.11. Dimensions of the sides of the (a) octahedra and (b) tetrahedra for the FCC, BCC, and hcp interstitial geometries.

5
Detailed Equations
for Various Crystal Systems

5.1. General Equations Applicable to Any System

Defining vectors: **a, b, c**
Plane indices : {hkl}
The lattice parameters are a, b, c; α, β, γ; in this case none of them are equal to each other.
Defining angles between basis vectors: α: b, c; β: a, c; γ: a, b.
Basis vectors: **a, b, c**: no convenient form with respect to **i , j, k** unit vectors.
Reciprocal lattice vectors:

$$\mathbf{a^*} = \frac{\mathbf{b \times c}}{V}, \quad \mathbf{b^*} = \frac{\mathbf{c \times a}}{V}, \quad \mathbf{c^*} = \frac{\mathbf{a \times b}}{V}. \tag{5.1}$$

Reciprocal lattice angles:

$$\cos(\alpha^*) = \frac{\cos\beta\cos\gamma - \cos\alpha}{\sin\gamma\sin\beta}, \tag{5.2}$$

$$\cos(\beta^*) = \frac{\cos\gamma\cos\alpha - \cos\beta}{\sin\gamma\sin\alpha}, \tag{5.3}$$

$$\cos(\gamma^*) = \frac{\cos\alpha\cos\beta - \cos\gamma}{\sin\alpha\sin\beta}. \tag{5.4}$$

Volume in direct space:

$$V = \mathbf{a \cdot b \times c} = abc\ \{1 - \cos^2\alpha - \cos^2\beta - \cos^2\gamma + 2\cos\alpha\cos\beta\cos\gamma\}^{1/2} \tag{5.5}$$

Volume in reciprocal space:

$$V^* = \frac{1}{V}. \tag{5.6}$$

Repeat spacing of points in a row along a direction:

$$\text{direction vector} = \mathbf{u} = [uvw] = u\,\mathbf{a} + v\,\mathbf{b} + w\,\mathbf{c}, \tag{5.7}$$

for u, v, w integers;

$$|\mathbf{u}| = \{a^2u^2 + b^2v^2 + c^2w^2 + 2\,b\,c\,v\,w\,\cos\alpha \\ + 2\,c\,a\,w\,u\,\cos\beta + 2\,a\,b\,u\,v\,\cos\gamma\}^{1/2}. \tag{5.8}$$

The repeat spacing in reciprocal space is given by

$$H = \frac{1}{|\mathbf{u}|} \cdot \tag{5.9}$$

Plane in reciprocal space:

$$\mathbf{g} = \{h\,k\,l\} = h\,\mathbf{a}^* + k\,\mathbf{b}^* + l\,\mathbf{c}^*, \quad |\mathbf{g}| = \frac{1}{d}; \tag{5.10}$$

d = distance between equivalent planes.
Magnitude of reciprocal lattice vectors:

$$|\mathbf{a}^*| = \frac{b\,c\,\sin\alpha}{V}, \quad |\mathbf{b}^*| = \frac{c\,a\,\sin\beta}{V}, \quad |\mathbf{c}^*| = \frac{a\,b\,\sin\gamma}{V}. \tag{5.11}$$

Spacing between equivalent planes

$$\frac{1}{d^2} = h^2\,a^{*2} + k^2\,b^{*2} + l^2\,c^{*2} + 2\,h\,k\,a^*\,b^*\,\cos\gamma^*$$
$$+ 2\,k\,l\,b^*\,c^*\,\cos\alpha^* + 2\,l\,h\,c^*\,a^*\,\cos\beta^*. \tag{5.12}$$

Angle between planes $(h_1\,k_1\,l_1)$ and $(h_2\,k_2\,l_2)$:

$$\cos\theta_p = \{h_1\,h_2\,|\mathbf{a}^*|^2 + k_1\,k_2\,|\mathbf{b}^*|^2 + l_1\,l_2\,|\mathbf{c}^*|^2$$

$$+ (h_1\,k_2 + h_2\,k_1)\,|\mathbf{a}^*|\,|\mathbf{b}^*|\,\cos\gamma^*$$

$$+ (k_1\,l_2 + k_2\,l_1)\,|\mathbf{b}^*|\,|\mathbf{c}^*|\,\cos\alpha^*$$

$$+ (l_1\,h_2 + l_2\,h_1)\,|\mathbf{c}^*|\,|\mathbf{a}^*|\,\cos\beta^*\}\,d_1 d_2. \tag{5.13}$$

Angle between directions $[u_1\,v_1\,w_1]$ and $[u_2\,v_2\,w_2]$:

$$\cos\theta_d = \{u_1 u_2 a^2 + v_1 v_2 b^2 + w_1 w_2 c^2 + (u_1 v_2 + u_2 v_1)\,a\,b\,\cos\gamma$$
$$+ (v_1 w_2 + v_2 w_1)\,b\,c\,\cos\alpha$$
$$+ (w_1 u_2 + w_2 u_1)\,c\,a\,\cos\beta\} \left(\frac{1}{|\mathbf{u}_1|\,|\mathbf{u}_2|}\right). \tag{5.14}$$

Relationship between direction and plane indices for the zero order Laue zone:

$$\frac{u}{k_1\,l_2 - k_2\,l_1} = \frac{v}{l_1\,h_2 - l_2\,h_1} = \frac{w}{h_1\,k_2 - h_2\,k_1}, \tag{5.15}$$

where $(h_1\,k_1\,l_1)$ and $(h_2\,k_2\,l_2)$ are two planes in the zone.
  Alternatively, if (hkl) and [uvw] are a plane and a direction, they are related in general by the following three expressions that are equal to each other:

$$\frac{u}{hb^2c^2\sin^2\alpha + kabc^2(\cos\alpha\cos\beta - \cos\gamma) + lab^2c(\cos\gamma\cos\alpha - \cos\beta)} =$$

$$\frac{v}{habc^2(\cos\alpha\cos\beta - \cos\gamma) + ka^2c^2\sin^2\beta + la^2bc(\cos\gamma\cos\beta - \cos\alpha)} =$$

$$\frac{w}{hab^2c(\cos\gamma\cos\alpha - \cos\beta) + ka^2bc(\cos\gamma\cos\beta - \cos\alpha) + la^2b^2\sin^2\gamma}$$

(5.16)

and also by the following equations:

$$\frac{h}{ua^2 + vab\cos\gamma + wca\cos\beta} = \frac{k}{vb^2 + uab\cos\gamma + wcb\cos\alpha}$$

$$= \frac{l}{wc^2 + vcb\cos\alpha + uca\cos\beta}$$

(5.17)

Angle between $\mathbf{u}^*$ and $\mathbf{u}$, where $\mathbf{u}^*$ is a zone axis such that $\mathbf{g} \cdot \mathbf{u}^* = 0$, and $\mathbf{u}$ is the direct lattice direction vector:

$$\cos\theta = \frac{\mathbf{u}^* \cdot \mathbf{u}}{|\mathbf{u}^*|\,\|\mathbf{u}|} = (u^*\mathbf{a}^* + b^*\mathbf{v}^* + w^*\mathbf{c}^*) \cdot (u\mathbf{a} + v\mathbf{b} + w\mathbf{c})$$
$$= u^*u\,\mathbf{a}^* \cdot \mathbf{a} + u^*v\,\mathbf{a}^* \cdot \mathbf{b} + u^*w\,\mathbf{a}^* \cdot \mathbf{c}$$
$$+ v^*u\,\mathbf{b}^* \cdot \mathbf{a} + v^*v\,\mathbf{b}^* \cdot \mathbf{b} + v^*w\,\mathbf{b}^* \cdot \mathbf{c}$$
$$+ w^*u\,\mathbf{c}^* \cdot \mathbf{a} + w^*v\,\mathbf{c}^* \cdot \mathbf{b} + w^*w\,\mathbf{c}^* \cdot \mathbf{c}.$$

(5.18)

For cubic, tetragonal, and orthorhombic systems, the mixed dot products are zero, leaving only the "diagonal" elements, which means that the direction vector and the reciprocal lattice vector are always parallel, or the zone axis is parallel to the normal to the plane in direct space. For hexagonal, trigonal, monoclinic and triclinic systems, the "off-diagonal" elements are not zero, meaning that the reciprocal lattice vector and the direct lattice vector of the normal to the plane are not parallel except in special cases.

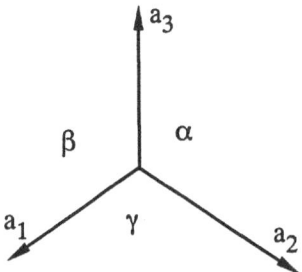

Figure 5.1. Direct space coordinate system.

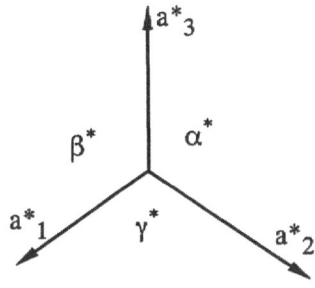

Figure 5.2. Reciprocal space coordinate system.

5.2. Cubic System

Defining vectors: $a_1 = a\,i$; $a_2 = a\,j$; $a_3 = a\,k$; i, j, k are unit vectors in an orthogonal right-handed coordinate system, and a is the lattice parameter. In the simple cubic lattice, a is the distance of closest approach $= a_{closest}$. In the FCC lattice a is $\sqrt{2}\ a_{closest}$ (along face diagonal). In the BCC lattice a is $\frac{2}{\sqrt{3}}\ a_{closest}$ (along body diagonal). See Figures 5.1 and 5.2 for the coordinate orientations.

Plane indices: (hkl) = reciprocal of intercepts along each of the basis vectors, and h, k, l each an integer. See Figures 5.3, 5.4, and 5.5 for lattice point locations. Reciprocal lattice vectors:

$$a^*_1 = \frac{a_2 \times a_3}{V}, \quad a^*_2 = \frac{a_3 \times a_1}{V}, \quad a^*_3 = \frac{a_1 \times a_2}{V}. \tag{5.19}$$

Various dot products:

$$a^*_1 \cdot a^*_1 = \frac{1}{a^2}, \quad a^*_2 \cdot a^*_1 = 0, \quad a^*_3 \cdot a^*_1 = 0,$$

$$a^*_1 \cdot a^*_2 = 0, \quad a^*_2 \cdot a^*_2 = \frac{1}{a^2}, \quad a^*_3 \cdot a^*_2 = 0,$$

$$a^*_1 \cdot a^*_3 = 0, \quad a^*_2 \cdot a^*_3 = 0, \quad a^*_3 \cdot a^*_3 = \frac{1}{a^2}. \tag{5.20}$$

Volume:

$$V = a_1 \cdot a_2 \times a_3 = a^3. \tag{5.21}$$

Repeat spacing in the direct lattice:

$$u = [uvw] = u\,a_1 + v\,a_2 + w\,a_3, \quad |u| = a\sqrt{u^2 + v^2 + w^2}. \tag{5.22}$$

A plane in reciprocal space is defined by

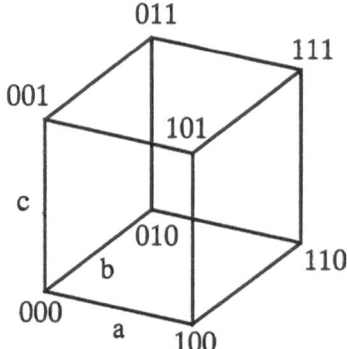

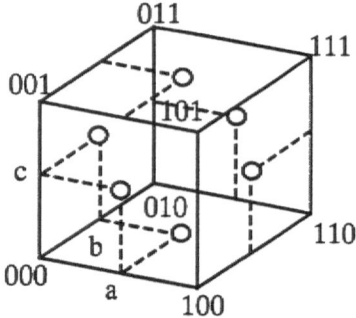

Figure 5.3. Simple cubic lattice with lattice points labeled.

Figure 5.4. Face-centered cubic lattice with lattice points labeled.

$$\mathbf{g} = (h\,k\,l) = h\,\mathbf{a^*}_1 + k\,\mathbf{a^*}_2 + l\,\mathbf{a^*}_3,\qquad(5.23)$$

$$|\mathbf{g}| = \frac{\sqrt{h^2 + k^2 + l^2}}{a} = \frac{1}{d},\qquad(5.24)$$

where d = distance between equivalent planes.
    Angle between planes:

$$\cos\theta_{12} = \frac{\mathbf{g}_1 \cdot \mathbf{g}_2}{|\mathbf{g}_1||\mathbf{g}_2|} = \frac{h_1h_2 + k_1k_2 + l_1l_2}{\sqrt{h_1^2 + k_1^2 + l_1^2}\sqrt{h_2^2 + k_2^2 + l_2^2}}.\qquad(5.25)$$

Angle between directions:

$$\cos\phi_{12} = \frac{\mathbf{u}_1 \cdot \mathbf{u}_2}{|\mathbf{u}_1\|\mathbf{u}_2|} = \frac{u_1u_2 + v_1v_2 + w_1w_2}{\sqrt{u_1^2 + v_1^2 + w_1^2}\sqrt{u_2^2 + v_2^2 + w_2^2}}.\qquad(5.26)$$

Relationship between indices for planes and for directions:

$$\frac{u}{h} = \frac{v}{k} = \frac{w}{l}.\qquad(5.27)$$

Zone law for higher order Laue zones:

$$\mathbf{g} \cdot \mathbf{u} = hu + kv + lw,\qquad(5.28)$$

$$\mathbf{g} \cdot \mathbf{u} = N,\qquad(5.29)$$

where N = the number of the Laue zone.
    Repeat spacing in reciprocal space:

$$H = \frac{1}{|\mathbf{u}|} = \frac{1}{a\sqrt{u^2 + v^2 + w^2}}.\qquad(5.30)$$

Atoms per unit cell for SC, FCC, and BCC lattices:

Lattice	Atoms/unit cell	Lattice Point Coordinates
SC	$1 = 8 \cdot 1/8$	[000],[100],[010],[001],[110],[111], [101],[011]
FCC	$4 = 8 \cdot 1/8 + 6 \cdot 1/2$	[000],[100],[010],[001],[110],[111], $[\frac{1}{2}0\frac{1}{2}]$, $[\frac{1}{2}\frac{1}{2}0]$ $[0\frac{1}{2}\frac{1}{2}]$ $[1\frac{1}{2}\frac{1}{2}]$,$[\frac{1}{2}1\frac{1}{2}]$,$[\frac{1}{2}\frac{1}{2}1]$
BCC	$2 = 8 \cdot 1/8 + 1$	[000],[100],[010],[001],[110],[111], [101],[011], $[\frac{1}{2}\frac{1}{2}\frac{1}{2}]$

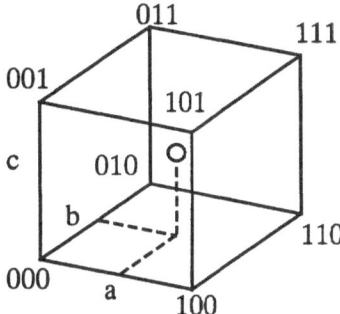

Figure 5.5. Body-centered cubic lattice with lattice points labeled.

5.3. Tetragonal System

Defining vectors: $a_1 = a\ i$; $a_2 = a\ j$; $a_3 = c\ k$; i, j, k are unit vectors in an orthogonal right-handed coordinate system, and a, c are the lattice parameters.   The c parameter is oriented along the 4-fold symmetry axis. The three axes are at 90° to each other. See Figures 5.6 and 5.7 for coordinate systems.
Plane indices: (hkl).
Reciprocal lattice vectors:

$$a^*_1 = \frac{a_2 \times a_3}{V} = \frac{i}{a}, \quad a^*_2 = \frac{a_3 \times a_1}{V} = \frac{j}{a}, \quad a^*_3 = \frac{a_1 \times a_2}{V} = \frac{k}{c}. \tag{5.34}$$

Various dot products:

$$
\begin{array}{lll}
a_1 \cdot a_1 = a^2, & a_1 \cdot a_2 = 0, & a_1 \cdot a_3 = 0, \\
a_2 \cdot a_1 = 0, & a_2 \cdot a_2 = a^2, & a_2 \cdot a_3 = 0, \\
a_3 \cdot a_1 = 0, & a_3 \cdot a_2 = 0, & a_3 \cdot a_3 = c^2.
\end{array}
\tag{5.35}
$$

$$
\begin{array}{lll}
a^*_1 \cdot a^*_1 = \dfrac{1}{a^2}, & a^*_1 \cdot a^*_2 = 0, & a^*_1 \cdot a^*_3 = 0, \\[2mm]
a^*_2 \cdot a^*_1 = 0, & a^*_2 \cdot a^*_2 = \dfrac{1}{a^2}, & a^*_2 \cdot a^*_3 = 0, \\[2mm]
a^*_3 \cdot a^*_1 = 0, & a^*_3 \cdot a^*_2 = 0, & a^*_3 \cdot a^*_3 = \dfrac{1}{c^2}.
\end{array}
\tag{5.36}
$$

$$
\begin{array}{lll}
a_1 \cdot a^*_1 = 1, & a_1 \cdot a^*_2 = 0, & a_1 \cdot a^*_3 = 0, \\
a_2 \cdot a^*_1 = 0, & a_2 \cdot a^*_2 = 1, & a_2 \cdot a^*_3 = 0, \\
a_3 \cdot a^*_1 = 0, & a_3 \cdot a^*_2 = 0, & a_3 \cdot a^*_3 = 1.
\end{array}
\tag{5.37}
$$

Volume:

$$V = a_1 \cdot a_2 \times a_3 = a^2\ c. \tag{5.38}$$

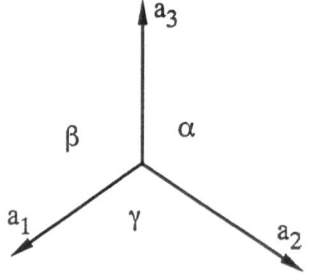

Figure 5.6. Direct space coordinate system.

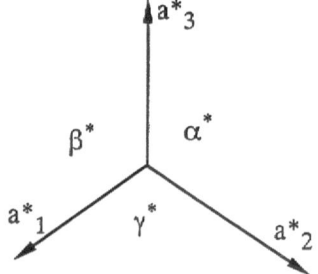

Figure 5.7. Reciprocal space coordinate system.

A direction in the direct lattice is defined by

$$\mathbf{u} = [uvw] = u\mathbf{a}_1 + v\mathbf{a}_2 + w\mathbf{a}_3, \tag{5.39}$$

$$|\mathbf{u}| = \sqrt{a^2(u^2 + v^2) + c^2 w^2} = a\sqrt{u^2 + v^2 + \gamma^2 w^2}, \tag{5.40}$$

where $\gamma = c/a$; u, v, w integers.

A plane in reciprocal space is defined by

$$\mathbf{g} = (hkl) = h\,\mathbf{a}^*_1 + k\,\mathbf{a}^*_2 + l\,\mathbf{a}^*_3, \tag{5.41}$$

$$|\mathbf{g}| = \sqrt{\frac{h^2 + k^2}{a^2} + \frac{l^2}{c^2}}, \tag{5.42}$$

$$|\mathbf{g}| = \frac{1}{a}\sqrt{h^2 + k^2 + \gamma^{-2} l^2} = \frac{1}{d}, \tag{5.43}$$

where d = distance between equivalent planes in the direct lattice.

Angle between planes:

$$\cos\theta_{12} = \frac{\mathbf{g}_1 \cdot \mathbf{g}_2}{|\mathbf{g}_1||\mathbf{g}_2|} = \frac{\dfrac{h_1 h_2 + k_1 k_2}{a^2} + \dfrac{l_1 l_2}{c^2}}{\sqrt{\dfrac{h_1^2 + k_1^2}{a^2} + \dfrac{l_1^2}{c^2}}\sqrt{\dfrac{h_2^2 + k_2^2}{a^2} + \dfrac{l_2^2}{c^2}}}$$

$$= \frac{h_1 h_2 + k_1 k_2 + \gamma^{-2} l_1 l_2}{\sqrt{h_1^2 + k_1^2 + \gamma^{-2} l_1^2}\sqrt{h_2^2 + k_2^2 + \gamma^{-2} l_2^2}}. \tag{5.44}$$

Angle between directions:

$$\cos\phi_{12} = \frac{\mathbf{u}_1 \cdot \mathbf{u}_2}{|\mathbf{u}_1||\mathbf{u}_2|} = \frac{(u_1 u_2 + v_1 v_2)a^2 + w_1 w_2 c^2}{\sqrt{(u_1^2 + v_1^2)a^2 + w_1^2 c^2}\sqrt{(u_2^2 + v_2^2)a^2 + w_2^2 c^2}}$$

$$= \frac{u_1 u_2 + v_1 v_2 + \gamma^2 w_1 w_2}{\sqrt{u_1^2 + v_1^2 + w_1^2 \gamma^2}\sqrt{u_2^2 + v_2^2 + w_2^2 \gamma^2}} \tag{5.45}$$

Relationships among the direction indices and the plane indices:

$$\frac{u}{h} = \frac{v}{k} = \frac{w\gamma^2}{l}. \tag{5.46}$$

Zone law for zero order Laue zone:

$$\mathbf{g} \cdot \mathbf{u} = hu + kv + lw = 0. \tag{5.47}$$

Zone law for higher order Laue zones:

$$\mathbf{g} \cdot \mathbf{u} = N, \tag{5.48}$$

where N = the number of the Laue zone.
    Repeat spacing in reciprocal space:

$$H = \frac{1}{|u|} = \frac{1}{a\sqrt{u^2 + v^2 + \gamma^2 w^2}}. \tag{5.49}$$

    In Figures 5.8-5.11 are shown the unit cells for the tetragonal system. Atoms per unit cell for simple tetragonal, face-centered tetragonal, body-centered tetragonal, and base-centered tetragonal lattices:

Lattice	Atoms/unit cell	Lattice Point Coordinates
Simple	$1 = 8 \cdot 1/8$	[000],[100],[010],[001],[110],[111], [101],[011]
Face-centered	$4 = 8 \cdot 1/8 + 6 \cdot 1/2$	[000],[100],[010],[001],[110],[111], $[101],[011],[\frac{1}{2}\,0\,\frac{1}{2}],[\frac{1}{2}\,\frac{1}{2}\,0],[0\,\frac{1}{2}\,\frac{1}{2}]$ $[\frac{1}{2}\,1\,\frac{1}{2}],[1\,\frac{1}{2}\,\frac{1}{2}],[\frac{1}{2}\,\frac{1}{2}\,1]$
Body centered	$2 = 8 \cdot 1/8 + 1$	[00],[100],[010],[001],[110],[111], $[101],[011],[\frac{1}{2}\,\frac{1}{2}\,\frac{1}{2}]$
Base-centered	$2 = 8 \cdot 1/8 + 2 \cdot 1/2$	[000],[100],[010],[001],[110],[111], $[101],[011],[\frac{1}{2}\,\frac{1}{2}\,0],[\frac{1}{2}\,\frac{1}{2}1]$

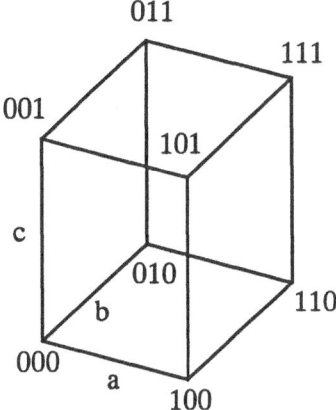

Figure 5.8. Simple tetragonal lattice.

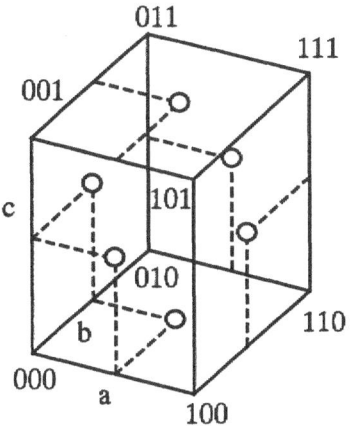

Figure 5.9. Face-centered tetragonal lattice. Note that such a lattice is not a Bravais lattice, but can be derived from a body-centered lattice.

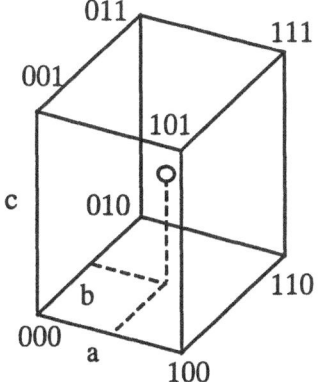

Figure 5.10. Body-centered lattice.

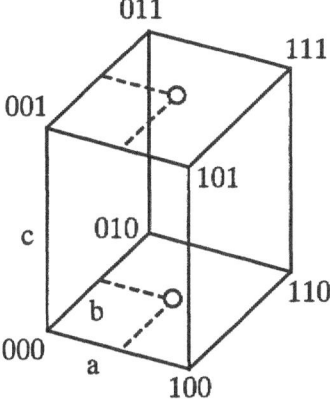

Figure 5.11. Base-centered lattice.

5.4. Orthorhombic System

Defining vectors: $a_1 = a\,i$; $a_2 = b\,j$; $a_3 = c\,k$; i, j, k are unit vectors in an orthogonal right-handed coordinate system, and a, b, c are the lattice parameters, oriented along the three 2-fold symmetry axes. The three axes are at 90° to each other and a, b, c not equal. See Figures 5.12 and 5.13.
  - Plane indices: {hkl}.

Reciprocal lattice vectors:

$$a^*_1 = \frac{a_2 \times a_3}{V} = \frac{i}{a}, \quad a^*_2 = \frac{a_3 \times a_1}{V} = \frac{j}{b}, \quad a^*_3 = \frac{a_1 \times a_2}{V} = \frac{k}{c}. \quad (5.50)$$

Various dot products:

$$
\begin{array}{lll}
a_1 \cdot a_1 = a^2, & a_1 \cdot a_2 = 0, & a_1 \cdot a_3 = 0, \\
a_2 \cdot a_1 = 0, & a_2 \cdot a_2 = b^2, & a_2 \cdot a_3 = 0, \\
a_3 \cdot a_1 = 0, & a_3 \cdot a_2 = 0, & a_3 \cdot a_3 = c^2.
\end{array} \quad (5.51)
$$

$$
\begin{array}{lll}
a^*_1 \cdot a^*_1 = \dfrac{1}{a^2}, & a^*_1 \cdot a^*_2 = 0, & a^*_1 \cdot a^*_3 = 0, \\[2mm]
a^*_2 \cdot a^*_1 = 0, & a^*_2 \cdot a^*_2 = \dfrac{1}{b^2}, & a^*_2 \cdot a^*_3 = 0, \\[2mm]
a^*_3 \cdot a^*_1 = 0, & a^*_3 \cdot a^*_2 = 0, & a^*_3 \cdot a^*_3 = \dfrac{1}{c^2}.
\end{array} \quad (5.52)
$$

$$
\begin{array}{lll}
a_1 \cdot a^*_1 = 1, & a_1 \cdot a^*_2 = 0, & a_1 \cdot a^*_3 = 0, \\
a_2 \cdot a^*_1 = 0, & a_2 \cdot a^*_2 = 1, & a_2 \cdot a^*_3 = 0, \\
a_3 \cdot a^*_1 = 0, & a_3 \cdot a^*_2 = 0, & a_3 \cdot a^*_3 = 1.
\end{array} \quad (5.53)
$$

Volume:

$$V = a_1 \cdot a_2 \times a_3 = abc. \quad (5.54)$$

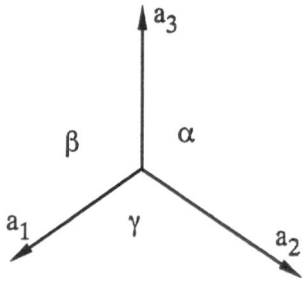

Figure 5.12. Direct space coordinate system.

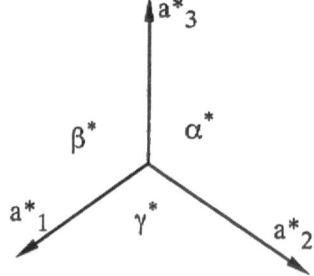

Figure 5.13. Reciprocal space coordinate system.

A direction in the direct lattice is defined by

$$\mathbf{u} = [uvw] = u\mathbf{a}_1 + v\mathbf{a}_2 + w\mathbf{a}_3, \tag{5.55}$$

$$|\mathbf{u}| = \sqrt{a^2u^2 + b^2v^2 + c^2w^2} = a\sqrt{u^2 + \delta^2v^2 + \gamma^2 w^2}, \tag{5.56}$$

where $\delta = b./a$ and $\gamma = c/a$; u, v, w integers.

A plane in reciprocal space is defined by

$$\mathbf{g} = (hkl) = h\,\mathbf{a}^*_1 + k\,\mathbf{a}^*_2 + l\,\mathbf{a}^*_3 , \tag{5.57}$$

$$|\mathbf{g}| = \sqrt{\frac{h^2}{a^2} + \frac{k^2}{b^2} + \frac{l^2}{c^2}}, \tag{5.58}$$

$$|\mathbf{g}| = \frac{1}{a}\sqrt{h^2 + \delta^{-2}k^2 + \gamma^{-2}l^2} = \frac{1}{d}, \tag{5.59}$$

where d = distance between equivalent planes in the direct lattice.

Angle between planes:

$$\cos\theta_{12} = \frac{\mathbf{g}_1\cdot\mathbf{g}_2}{|\mathbf{g}_1||\mathbf{g}_2|} = \frac{\dfrac{h_1h_2}{a^2} + \dfrac{k_1k_2}{b^2} + \dfrac{l_1l_2}{c^2}}{\sqrt{\dfrac{h_1^2}{a^2} + \dfrac{k_1^2}{b^2} + \dfrac{l_1^2}{c^2}}\sqrt{\dfrac{h_2^2}{a^2} + \dfrac{k_2^2}{b^2} + \dfrac{l_2^2}{c^2}}}$$

$$= \frac{h_1h_2 + \delta^{-2}k_1k_2 + \gamma^{-2}l_1l_2}{\sqrt{h_1^2 + \delta^{-2}k_1^2 + \gamma^{-2}l_1^2}\sqrt{h_2^2 + \delta^{-2}k_2^2 + \gamma^{-2}l_2^2}}. \tag{5.60}$$

Angle between directions:

$$\cos\phi_{12} = \frac{\mathbf{u}_1\cdot\mathbf{u}_2}{|\mathbf{u}_1||\mathbf{u}_2|} = \frac{u_1u_2a^2 + v_1v_2b^2 + w_1w_2c^2}{\sqrt{u_1^2a^2 + v_1^2b^2 + w_1^2c^2}\sqrt{u_2^2a^2 + v_2^2b^2 + w_2^2c^2}}$$

$$= \frac{u_1u_2 + \delta^2v_1v_2 + \gamma^2w_1w_2}{\sqrt{u_1^2 + v_1^2\delta^2 + w_1^2\gamma^2}\sqrt{u_2^2 + v_2^2\delta^2 + w_2^2\gamma^2}}. \tag{5.61}$$

Relationships among the direction indices and the plane indices:

$$\frac{ua^2}{h} = \frac{vb^2}{k} = \frac{wc^2}{l}, \quad \frac{u}{h} = \frac{v\delta^2}{k} = \frac{w\gamma^2}{l}. \tag{5.62}$$

Zone law for zero order Laue zone:

$$\mathbf{g}\cdot\mathbf{u} = hu + kv + lw = 0. \tag{5.63}$$

Zone law for higher order Laue zones:

$$\mathbf{g}\cdot\mathbf{u} = N, \tag{5.64}$$

where N = the number of the Laue zone.

Repeat spacing in reciprocal space:

$$H = \frac{1}{|\mathbf{u}|} = \frac{1}{a\sqrt{u^2 + \delta^2 v^2 + \gamma^2 w^2}} . \tag{5.65}$$

In Figures 5.14-5.17 are shown the unit cells for orthorhombic. Atoms per unit cell for simple orthorhombic, face-centered orthorhombic, body-centered orthorhombic and base-centered orthorhombic lattices are shown in the table below:

Lattice	Atoms/unit cell	Lattice Point Coordinates
Simple	$1 = 8 \cdot 1/8$	[000], [100], [010], [001], [110], [111], [101],[011]
Face-centered	$4 = 8 \cdot 1/8 + 6 \cdot 1/2$	[000],[100],[010],[001],[110],[111], [101],[011],$[\frac{1}{2}0\frac{1}{2}]$,$[\frac{1}{2}\frac{1}{2}0]$,$[0\frac{1}{2}\frac{1}{2}]$, $[1\frac{1}{2}\frac{1}{2}]$,$[\frac{1}{2}1\frac{1}{2}]$,$[\frac{1}{2}\frac{1}{2}1]$
Body-centered	$2 = 8 \cdot 1/8 + 1$	[000],[100],[010],[001],[110],[111], [101],[011],$[\frac{1}{2}\frac{1}{2}\frac{1}{2}]$
Base-centered	$2 = 8 \cdot 1/8 + 2 \cdot 1/2$	[000],[100],[010],[001],[110],[111], [101],[011],$[\frac{1}{2}\frac{1}{2}0]$,$[\frac{1}{2}\frac{1}{2}1]$

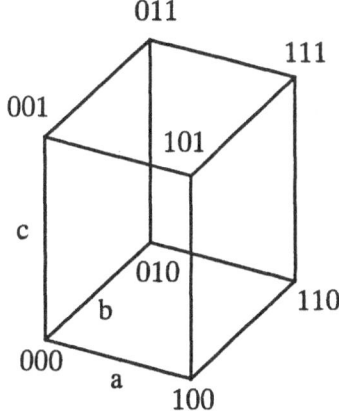

Figure 5.14. Simple orthorhombic lattice.

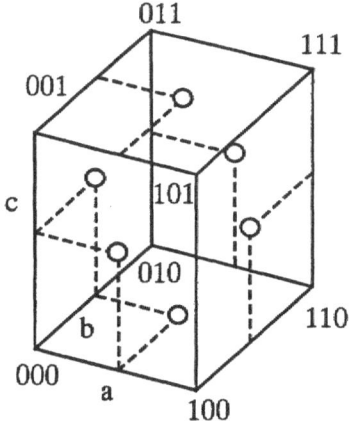

Figure 5.15. Face-centered orthorhombic lattice.

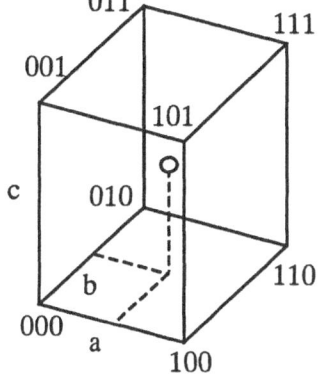

Figure 5.16. Body-centered orthorhombic lattice.

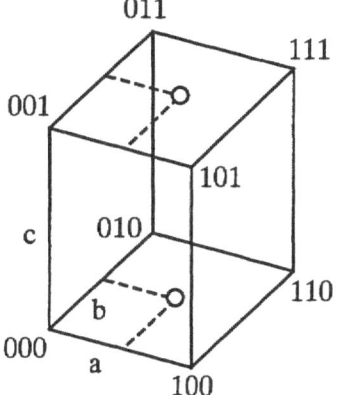

Figure 5.17. Base-centered orthorhombic lattice.

5.5. Monoclinic System (b Axis Unique)

Defining vectors:

$$a_1 = a\,i, \quad a_2 = b\,j, \quad a_3 = c\,k' = -c\cos\beta'\,i + c\sin\beta'\,k, \qquad (5.66)$$

where i, j, k are unit vectors in an orthogonal right-handed coordinate system and a, b, c are the lattice parameters and they are not equal to each other. The axes a and b and the axes b and c are at 90° to each other. The angle between a and c is not 90°. As shown in Figures 5.18 and 5.19, 90°< β <120°, i.e., $\beta' = \pi - \beta$. In terms of β, the third basis vector a_3 is written

$$a_3 = c\,k' = -c\cos(\pi-\beta)\,i + c\sin(\pi-\beta)\,k. \qquad (5.67)$$

For reference purposes, note that

$$\cos(\pi-\beta) = \cos\pi\cos\beta + \sin\pi\sin\beta = -\cos\beta, \qquad (5.68)$$

$$\sin(\pi-\beta) = \sin\pi\cos\beta - \cos\pi\sin\beta = \sin\beta. \qquad (5.69)$$

Hence, a_3 is written

$$a_3 = c\,k' = -c\cos\beta'\,i + c\sin\beta'\,k. \qquad (5.70)$$

Plane indices: (hkl).

Reciprocal lattice vectors:

$$a^*_1 = \frac{a_2 \times a_3}{V} = \frac{i}{a} - \frac{\cos\beta}{a\sin\beta}\,k,$$

$$a^*_2 = \frac{a_3 \times a_1}{V} = \frac{j}{b},$$

$$a^*_3 = \frac{a_1 \times a_2}{V} = \frac{1}{c\sin\beta}\,k. \qquad (5.71)$$

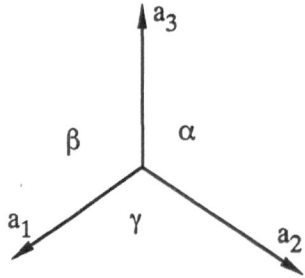

Figure 5.18. Direct space coordinate system.

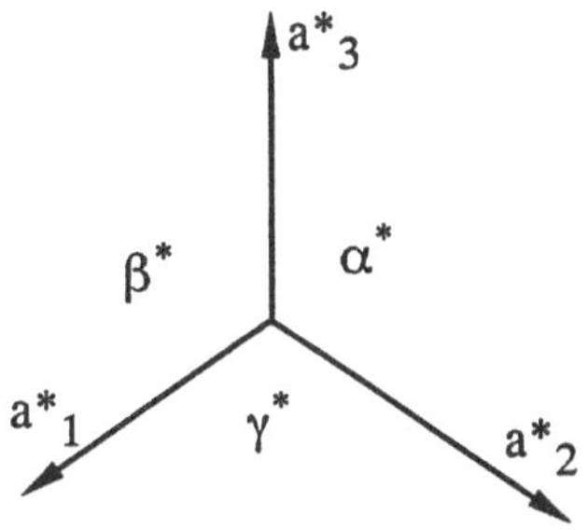

Figure 5.19. Reciprocal space coordinate system.

Various dot products:

$$a_1 \cdot a_1 = a^2, \qquad a_2 \cdot a_1 = 0, \qquad a_3 \cdot a_1 = a\,c\,\cos\beta,$$

$$a_1 \cdot a_2 = 0, \qquad a_2 \cdot a_2 = b^2, \qquad a_3 \cdot a_2 = 0,$$

$$a_1 \cdot a_3 = a\,c\,\cos\beta, \qquad a_2 \cdot a_3 = 0, \qquad a_3 \cdot a_3 = c^2. \tag{5.72}$$

Reciprocal lattice dot products:

$$a^*_1 \cdot a^*_1 = \frac{1}{a^2} + \frac{\cos^2\beta}{a^2\sin^2\beta}, \qquad a^*_2 \cdot a^*_1 = 0, \qquad a^*_3 \cdot a^*_1 = -\frac{\cos\beta}{a\,c\,\sin^2\beta},$$

$$= \frac{1}{a^2\sin^2\beta}$$

$$a^*_1 \cdot a^*_2 = 0, \qquad\qquad a^*_2 \cdot a^*_2 = \frac{1}{b^2}, \qquad a^*_3 \cdot a^*_2 = 0,$$

$$a^*_1 \cdot a^*_3 = -\frac{\cos\beta}{a\,c\,\sin^2\beta}, \qquad a^*_2 \cdot a^*_3 = 0, \qquad a^*_3 \cdot a^*_3 = \frac{1}{c^2\sin^2\beta}.$$

$$\tag{5.73}$$

Mixed dot products:

$$
\begin{array}{lll}
a_1 \cdot a^*_1 = 1, & a_1 \cdot a^*_2 = 0, & a_1 \cdot a^*_3 = 0, \\
a_2 \cdot a^*_1 = 0, & a_2 \cdot a^*_2 = 1, & a_2 \cdot a^*_3 = 0, \\
a_3 \cdot a^*_1 = 0, & a_3 \cdot a^*_2 = 0, & a_3 \cdot a^*_3 = 1.
\end{array}
\tag{5.74}
$$

Volume:

$$V = a_1 \cdot a_2 \times a_3 = abc\,\sin\beta. \tag{5.75}$$

A direction in the direct lattice is defined by

$$u = [uvw] = ua_1 + va_2 + wa_3, \tag{5.76}$$

$$|u| = \sqrt{a^2u^2 + b^2v^2 + c^2w^2 + 2uwac\,\cos\beta}. \tag{5.77}$$

In converting from direct space to reciprocal space, the angle, β^*, between $\mathbf{a}^*_1$ and $\mathbf{a}^*_3$ is found in reciprocal space to be related to β by

$$\cos \beta^* = -\cos \beta. \tag{5.78}$$

See Figure 5.19. Hence, for the monoclinic system, the direct lattice angles and the reciprocal lattice angles are related as

$$\beta^* = \pi - \beta, \quad \alpha^* = \alpha, \quad \gamma^* = \gamma. \tag{5.79}$$

A plane in reciprocal space is defined by

$$\mathbf{g} = (hkl) = h\,\mathbf{a}^*_1 + k\,\mathbf{a}^*_2 + l\,\mathbf{a}^*_3, \tag{5.80}$$

$$|\mathbf{g}| = \sqrt{\frac{h^2}{a^2 \sin^2 \beta} + \frac{k^2}{b^2} + \frac{l^2}{c^2 \sin^2 \beta} + \frac{2hl \cos \beta^*}{ac \sin^2 \beta}} = \frac{1}{d}, \tag{5.81}$$

where d = distance between equivalent planes in the direct lattice.

Angle between planes:

$$\cos \theta_{12} = \frac{\mathbf{g}_1 \cdot \mathbf{g}_2}{|\mathbf{g}_1||\mathbf{g}_2|} = \frac{\dfrac{h_1 h_2}{a^2 \sin^2 \beta} + \dfrac{k_1 k_2}{b^2} + \dfrac{l_1 l_2}{c^2 \sin^2 \beta} + \dfrac{(l_1 h_2 + l_2 h_1) \cos \beta^*}{ac \sin^2 \beta}}{\dfrac{1}{d_1 d_2}},$$

$$\tag{5.82}$$

$$\frac{1}{d_i} = \sqrt{\frac{h_i^2}{a^2 \sin^2 \beta} + \frac{k_i^2}{b^2} + \frac{l_i^2}{c^2 \sin^2 \beta} + \frac{2h_i l_i \cos \beta^*}{ac \sin^2 \beta}}, \tag{5.83}$$

Angle between directions:

$$\cos \phi_{12} = \frac{\mathbf{u}_1 \cdot \mathbf{u}_2}{|\mathbf{u}_1||\mathbf{u}_2|} = \frac{u_1 u_2 a^2 + v_1 v_2 b^2 + w_1 w_2 c^2 + ac\,(w_1 u_2 + w_2 u_1)}{|\mathbf{u}_1||\mathbf{u}_2|}. \tag{5.84}$$

Relationships among the direction indices and the plane indices:

$$\frac{u}{hb^2 c^2 - lab^2 c \cos \beta} = \frac{v}{kc^2 a^2 \sin^2 \beta} = \frac{w}{la^2 b^2 - hab^2 c \cos \beta}. \tag{5.85}$$

Zone law for zero order Laue zone:
$$\mathbf{g} \cdot \mathbf{u} = hu + kv + lw + 0. \tag{5.86}$$

Zone law for higher order Laue zones:

$$\mathbf{g} \cdot \mathbf{u} = N, \tag{5.87}$$

where N = the number of the Laue zone.
Repeat spacing in reciprocal space:

$$H = \frac{1}{|\mathbf{u}|} = \frac{1}{\sqrt{a^2u^2 + b^2v^2 + c^2w^2 + 2uwac\cos\beta}}. \tag{5.88}$$

The unit cells are shown in Figures 5.20-5.22. Atoms per unit cell for simple monoclinic and base-centered monoclinic lattices are shown in the table below:

Lattice	Atoms/unit cell	Lattice Point Coordinates
Simple	$1 = 8 \bullet 1/8$	[000],[100],[010],[001],[110],[111], [101],[011]
Base-centered	$2 = 8 \bullet 1/8 + 2 \bullet 1/2$	[000],[100],[010],[001],[110],[111], [101],[011], $[\frac{1}{2}\frac{1}{2}0],[\frac{1}{2}\frac{1}{2}1]$

Note that the actual distances for the indices are to be multiplied by [a,b,c].

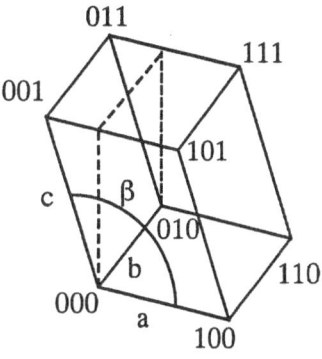

Figure 5.20. Simple monoclinic lattice.

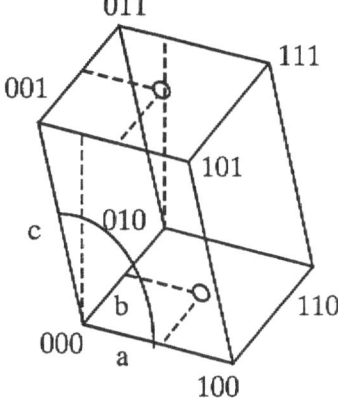

Figure 5.21. Base-centered monoclinic lattice.

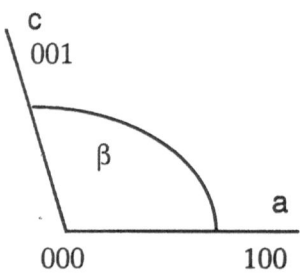

Figure 5.22. Projection along [010] axis for b-unique definition of axes.

5.6. Triclinic System

The triclinic system is the same as the general case given in Section 5.1. The unit cell is illustrated in Figure 5.23.

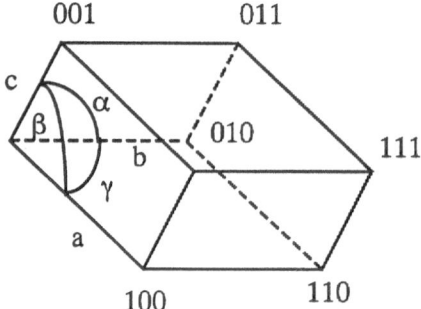

Figure 5.23. Triclinic unit cell:a ≠ b ≠ c;
α ≠ β ≠ γ.

5.7. Trigonal System

5.7.1. Basis vectors

The trigonal system may be referenced to two sets of basis vectors. The first set is a primitive trigonal set, in which the lattice translation parameters are all equal and which are at equal angles to each other. Hence, in this basis a length, a, and an angle, α, are required to characterize the unit cell (Figure 5.24). The alternate reference is in terms of a hexagonal basis system with lattice parameters a_H and c_H. The reciprocal lattice coordinate system is shown in Figure 5.25.

The two reference vector systems are related via the following equations. Note that the hexagonal basis is the three-axis, three-index basis (3, 3), explained in detail in Section 5.8.

For a_1 = hcp basis, a_2 = hcp basis, c_H = hcp basis, the trigonal basis vectors are

$$\mathbf{a} = \frac{2}{3}\mathbf{a}_1 + \frac{1}{3}\mathbf{a}_2 + \frac{1}{3}\mathbf{c}_H, \tag{5.89}$$

$$\mathbf{b} = -\frac{1}{3}\mathbf{a}_1 + \frac{1}{3}\mathbf{a}_2 + \frac{1}{3}\mathbf{c}_H, \tag{5.90}$$

$$\mathbf{c} = -\frac{1}{3}\mathbf{a}_1 - \frac{2}{3}\mathbf{a}_2 + \frac{1}{3}\mathbf{c}_H. \tag{5.91}$$

This choice of defining vectors is the preferred set, since it conforms to the *obverse* setting listed in the International Crystallography Tables [ITC, 1983], i.e., the setting in which the points in the trigonal cell are given in terms of **hexagonal** lattice points, defined by 0, 0, 0; 2/3, 1/3, 1/3; 1/3, 2/3, 2/3 in the hexagonal reference system. OTHER CHOICES OF BASIS VECTORS ARE IN USE. CHECK CAREFULLY TO BE CERTAIN WHICH BASIS VECTOR SET IS BEING USED.

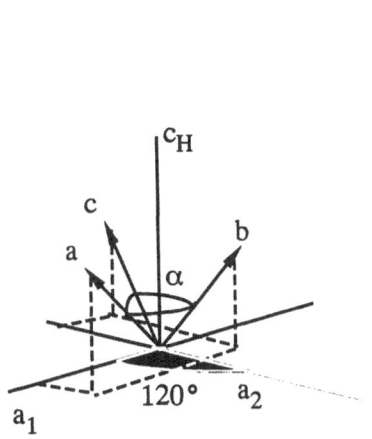

Figure 5.24. Coordinate system for trigonal in direct space.

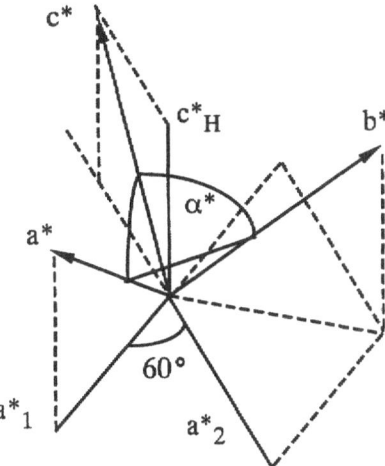

Figure 5.25. Coordinate system for trigonal in reciprocal space.

5.7.2. Various equations

Magnitude of the **trigonal a, b, c** vectors is, for $|\,a_1\,| = |\,a_2\,| = a_H$,

$$|\,\mathbf{a}\,| = |\,\mathbf{b}\,| = |\,\mathbf{c}\,| = \frac{1}{3}\sqrt{3\,a_H^2 + c_H^2}. \tag{5.92}$$

Magnitudes of the hexagonal lattice parameters in terms of the trigonal a, α are given by

$$a_H = a\sqrt{2(1 - \cos\alpha)}; \quad c_H = a\sqrt{3(1 + 2\cos\alpha)}. \tag{5.93}$$

The angle, α, between trigonal axes is

$$\cos\alpha = \frac{\mathbf{a}\cdot\mathbf{b}}{|\mathbf{a}||\mathbf{b}|} = \frac{-\dfrac{3a_H^2}{2} + c_H^2}{3a_H^2 + c_H^2} = \frac{2\gamma^2 - 3}{2\gamma^2 + 6}; \tag{5.94}$$

$$\sin\frac{\alpha}{2} = \frac{3}{2\sqrt{3 + \left(\dfrac{c_H}{a_H}\right)^2}}, \tag{5.95}$$

where $\gamma = c_H / a_H$.

Various dot products:

$$\mathbf{a}\cdot\mathbf{b} = \mathbf{b}\cdot\mathbf{c} = \mathbf{c}\cdot\mathbf{a} = -\frac{a_H^2}{6} + \frac{c^2}{9}. \tag{5.96}$$

Dot products in terms of rhombohedral lattice parameters are

$$\mathbf{a}\cdot\mathbf{a} = a^2, \qquad \mathbf{b}\cdot\mathbf{a} = a^2\cos\alpha, \qquad \mathbf{c}\cdot\mathbf{a} = a^2\cos\alpha,$$

$$\mathbf{a}\cdot\mathbf{b} = a^2\cos\alpha, \qquad \mathbf{b}\cdot\mathbf{b} = a^2, \qquad \mathbf{c}\cdot\mathbf{b} = a^2\cos\alpha,$$

$$\mathbf{a}\cdot\mathbf{c} = a^2\cos\alpha, \qquad \mathbf{b}\cdot\mathbf{c} = a^2\cos\alpha, \qquad \mathbf{c}\cdot\mathbf{c} = a^2. \tag{5.97}$$

From Section 5.8, we note that the volume in the hexagonal reference is

$$V_H = \frac{a_H^2 c_H \sqrt{3}}{2}, \tag{5.98}$$

and volume in the trigonal reference is

$$V_R = \frac{V_H}{3}. \tag{5.99}$$

Cross products among the trigonal basis vectors are, in terms of the hexagonal basis vectors:

$$\mathbf{a} \times \mathbf{b} = \frac{V_H}{3c_H{}^2}\,c_H - \frac{c_H}{3\sqrt{3}}\,[a_1 + a_2], \tag{5.100}$$

$$\mathbf{b} \times \mathbf{c} = \frac{V_H}{3c_H{}^2}\,c_H + \frac{c_H}{3\sqrt{3}}\,[2a_1 + a_2], \tag{5.101}$$

$$\mathbf{c} \times \mathbf{a} = \frac{V_H}{3c_H{}^2}\,c_H + \frac{c_H}{3\sqrt{3}}\,[-a_1 + a_2]. \tag{5.102}$$

Trigonal reciprocal lattice vectors in terms of the hexagonal basis vectors are

$$\mathbf{a^*} = \frac{\mathbf{b} \times \mathbf{c}}{V_R} = \frac{1}{c_H{}^2}\,c_H + \frac{2}{3a_H{}^2}\,(2a_1 + a_2) = c^*{}_H + a^*{}_1, \tag{5.103}$$

$$\mathbf{b^*} = \frac{\mathbf{c} \times \mathbf{a}}{V_R} = \frac{1}{c_H{}^2}\,c_H + \frac{2}{3a_H{}^2}\,(-a_1 + a_2) = c^*{}_H - a^*{}_1 + a^*{}_2, \tag{5.104}$$

$$\mathbf{c^*} = \frac{\mathbf{a} \times \mathbf{b}}{V_R} = \frac{1}{c_H{}^2}\,c_H - \frac{2}{3a_H{}^2}\,(a_1 + 2a_2) = c^*{}_H - a^*{}_2. \tag{5.105}$$

Magnitude of the trigonal reciprocal lattice vectors is

$$|\mathbf{a^*}| = |\mathbf{b^*}| = |\mathbf{c^*}| = \sqrt{\frac{1}{c_H{}^2} + \frac{4}{3a_H{}^2}}. \tag{5.106}$$

Mixed dot products between direct and reciprocal lattice vectors are

$$\mathbf{a^*} \cdot \mathbf{a} = 1, \qquad \mathbf{b^*} \cdot \mathbf{a} = 0, \qquad \mathbf{c^*} \cdot \mathbf{a} = 0,$$

$$\mathbf{a^*} \cdot \mathbf{b} = 0, \qquad \mathbf{b^*} \cdot \mathbf{b} = 1, \qquad \mathbf{c^*} \cdot \mathbf{b} = 0,$$

$$\mathbf{a^*} \cdot \mathbf{c} = 0, \qquad \mathbf{b^*} \cdot \mathbf{c} = 0, \qquad \mathbf{c^*} \cdot \mathbf{c} = 1. \tag{5.107}$$

Dot products among the reciprocal lattice vectors are

$$\mathbf{a^*} \cdot \mathbf{b^*} = \mathbf{b^*} \cdot \mathbf{c^*} = \mathbf{c^*} \cdot \mathbf{a^*} = |\mathbf{a^*}|^2 \cos \alpha^* = \frac{1}{c_H{}^2} - \frac{2}{3a_H{}^2}, \tag{5.108}$$

and for reciprocal lattice vectors referenced to the rhombohedral parameters:

$$\mathbf{a^*} \cdot \mathbf{a^*} = \mathbf{c^*} \cdot \mathbf{c^*} = \frac{1}{3a^2}\left(\frac{4 + 5\cos\alpha}{\cos\alpha - \cos 2\alpha}\right), \tag{5.109}$$

$$\mathbf{b^*} \cdot \mathbf{b^*} = \frac{1}{a^2}\left(\frac{1 + \cos\alpha}{\cos\alpha - \cos 2\alpha}\right), \tag{5.110}$$

$$\mathbf{a^*} \cdot \mathbf{b^*} = \mathbf{b^*} \cdot \mathbf{c^*} = \frac{1}{a^2}\left(\frac{1 - \sqrt{3} - (2\sqrt{3} + 1)\cos\alpha}{\cos\alpha - \cos 2\alpha}\right), \tag{5.111}$$

$$\mathbf{a^*} \cdot \mathbf{c^*} = \frac{1}{3a^2(1 + 2\cos\alpha)}. \tag{5.112}$$

A plane in reciprocal space is defined by

$$\mathbf{g} = h\mathbf{a^*} + k\mathbf{b^*} + l\mathbf{c^*}, \tag{5.113}$$

$$|\mathbf{g}| = |\mathbf{a^*}|\sqrt{h^2 + k^2 + l^2 + 2(hk + kl + lh)\cos\alpha^*} = \frac{1}{d}, \tag{5.114}$$

$$\cos\alpha^* = \frac{3 - 2\gamma^2}{3 + 4\gamma^2}. \tag{5.115}$$

Angle between planes is

$$\cos\theta_{ij} = \frac{\mathbf{g_i} \cdot \mathbf{g_j}}{|\mathbf{g_i}||\mathbf{g_j}|}, \tag{5.116}$$

where

$$\mathbf{g_i} \cdot \mathbf{g_j} = |\mathbf{a^*}|^2 [h_i h_j + k_i k_j + l_i l_j$$
$$+ (h_i k_j + h_i l_j + k_i h_j + k_i l_j + l_i h_j + l_i k_j)\cos\alpha^*]. \tag{5.117}$$

Angle between two directions is

$$\cos\phi_{ij} = \frac{\mathbf{u_i} \cdot \mathbf{u_j}}{|\mathbf{u_i}||\mathbf{u_j}|}, \tag{5.118}$$

$$\mathbf{u_i} = u_i\mathbf{a} + v_i\mathbf{b} + w_i\mathbf{c}, \tag{5.119}$$

and

$$\mathbf{u_j} = u_j\mathbf{a} + v_j\mathbf{b} + w_j\mathbf{c}, \tag{5.120}$$

$$|\mathbf{u}| = |\mathbf{a}|\sqrt{u^2 + v^2 + w^2 + 2(uv + vw + wu)\cos\alpha}, \tag{5.121}$$

$$\mathbf{u_i} \cdot \mathbf{u_j} = |\mathbf{a}|^2 [u_i u_j + v_i v_j + w_i w_j$$
$$+ (u_i v_j + u_i w_j + v_i u_j + v_i w_j + w_i u_j + w_i v_j)\cos\alpha], \tag{5.122}$$

where the u, v, w are the **trigonal** direction indices.

Given $<U_H V_H W_H>$ in (3,3) notation. To convert to trigonal indices, use the equations

$$u_T = U_H + W_H, \quad v_T = -U_H + V_H + W_H, \quad w_T = -V_H + W_H. \tag{5.123}$$

Given $<u_T\ v_T\ w_T>$, to find the hexagonal indices in (3, 3) notation use the equations

$$U_H = \frac{2}{3}u_T - \frac{1}{3}v_T - \frac{1}{3}w_T,$$

$$V_H = \frac{1}{3}u_T + \frac{1}{3}v_T - \frac{2}{3}w_T,$$

$$W_H = \frac{1}{3}u_T + \frac{1}{3}v_T + \frac{1}{3}w_T. \tag{5.124}$$

These conversion equations are obtained by using the transformation matrix (see ITC, 1983, p. 70-79), which relates the two basis vector systems:

$$\begin{pmatrix} u_T \\ v_T \\ w_T \end{pmatrix} = \begin{pmatrix} 1 & 0 & 1 \\ -1 & 1 & 1 \\ 0 & -1 & 1 \end{pmatrix} \begin{pmatrix} U_H \\ V_H \\ W_H \end{pmatrix}, \tag{5.125}$$

$$\begin{pmatrix} U_H \\ V_H \\ W_H \end{pmatrix} = \begin{pmatrix} \frac{2}{3} & -\frac{1}{3} & -\frac{1}{3} \\ \frac{1}{3} & \frac{1}{3} & -\frac{2}{3} \\ \frac{1}{3} & \frac{1}{3} & \frac{1}{3} \end{pmatrix} \begin{pmatrix} u_T \\ v_T \\ w_T \end{pmatrix}. \tag{5.126}$$

Relationships among the plane indices (see Table 5.1 for list of directions and planes):

(a) given $(h_T\ k_T\ l_T)$; rhombohedral to hcp:

$$H_H = h_T - k_T,$$
$$K_H = k_T - l_T,$$
$$L_H = h_T + k_T + l_T. \tag{5.127}$$

(b) given $(H_H\ K_H\ L_H)$; hcp to rhombohedral:

$$h_T = \frac{2}{3}H_H + \frac{1}{3}K_H + \frac{1}{3}L_H,$$

$$k_T = -\frac{1}{3}H_H + \frac{1}{3}K_H + \frac{1}{3}L_H,$$

$$l_T = -\frac{1}{3}H_H - \frac{2}{3}K_H + \frac{1}{3}L_H. \tag{5.128}$$

Table 5.1. Trigonal Indices in Rhombohedral and Hexagonal Bases.

Directions				Planes			
Rhomb.	Hexagonal			Rhomb.	Hexagonal		
	(4, 4)	(3, 3)	(O-H)		(4, 4)	(3, 3)	(O-H)
[111]	[0001]	[001]	[001]	(111)	(0001)	(001)	(001)
$30\bar{3}$	$11\bar{2}0$	330	330	$10\bar{1}$	$11\bar{2}0$	110	130
$03\bar{3}$	$\bar{1}2\bar{1}0$	030	$1/2[\bar{3}30]$	$01\bar{1}$	$\bar{1}2\bar{1}0$	$\bar{1}20$	$\bar{1}30$
$\bar{3}30$	$2\bar{1}\bar{1}0$	300	300	$1\bar{1}0$	$2\bar{1}\bar{1}0$	$\bar{2}10$	200
$14\bar{2}$	$\bar{1}2\bar{1}1$	031	$1/2[\bar{3}32]$	$14\bar{2}$	$\bar{1}2\bar{1}1$	$\bar{1}21$	$\bar{1}31$
693	$\bar{1}2\bar{1}6$	036	$1/2[\bar{1}\bar{1}4]$	231	$\bar{1}2\bar{1}6$	$\bar{1}26$	$\bar{1}36$
630	$11\bar{2}3$	333	$1/2[336]$	210	$11\bar{2}3$	113	133
360	$\bar{1}2\bar{1}3$	033	$1/2[\bar{3}36]$	120	$\bar{1}2\bar{1}3$	$\bar{1}23$	$\bar{1}33$
603	$\bar{2}113$	303	303	021	$\bar{2}113$	$\bar{2}13$	$\bar{2}03$
$39\bar{3}$	$\bar{2}\bar{4}23$	063	$\bar{1}11$	$13\bar{1}$	$\bar{2}\bar{4}23$	$\bar{2}43$	$\bar{2}63$
$2\bar{1}\bar{1}$	$10\bar{1}0$	210	$1/2[310]$	$2\bar{1}\bar{1}$	$10\bar{1}0$	100	110
$11\bar{2}$	$01\bar{1}0$	120	010	$11\bar{2}$	$01\bar{1}0$	010	020
$\bar{1}2\bar{1}$	$\bar{1}100$	$\bar{1}10$	$1/2[\bar{3}10]$	$\bar{1}2\bar{1}$	$\bar{1}100$	$\bar{1}10$	$\bar{1}10$
300	$10\bar{1}1$	211	$1/2[312]$	100	$10\bar{1}1$	101	111
$00\bar{3}$	$01\bar{1}\bar{1}$	$12\bar{1}$	$01\bar{1}$	$00\bar{1}$	$01\bar{1}1$	$01\bar{1}$	$02\bar{1}$
030	$\bar{1}101$	$\bar{1}11$	$1/2[\bar{3}12]$	010	$\bar{1}101$	$\bar{1}11$	$\bar{1}11$
711	$20\bar{2}3$	423	313	711	$20\bar{2}3$	203	223
$55\bar{1}$	$02\bar{2}3$	243	023	$55\bar{1}$	$02\bar{2}3$	023	043
171	$\bar{2}203$	$\bar{2}23$	$\bar{3}13$	171	$\bar{2}203$	$\bar{2}23$	$\bar{2}23$
411	$10\bar{1}2$	212	$1/2[314]$	411	$10\bar{1}2$	102	112
330	$01\bar{1}2$	122	012	110	$01\bar{1}2$	012	022
303	$\bar{1}102$	$\bar{1}12$	$1/2[3\bar{1}4]$	101	$\bar{1}102$	$\bar{1}12$	$\bar{1}12$
$41\bar{5}$	$12\bar{3}0$	450	$1/2[350]$	$41\bar{5}$	$12\bar{3}0$	120	150
$96\bar{3}$	$5\bar{1}\bar{4}0$	930	$1/2[510]$	$3\bar{2}\bar{1}$	$5\bar{1}\bar{4}0$	$5\bar{1}0$	530
$69\bar{3}$	$\bar{5}4\bar{1}0$	$6\bar{3}0$	$1/2[5\bar{1}0]$	$23\bar{1}$	$\bar{5}4\bar{1}0$	$5\bar{4}0$	$5\bar{3}0$
$12,\bar{3},0$	$5\bar{1}\bar{4}3$	933	$1/2[512]$	$4\bar{1}0$	$5\bar{1}\bar{4}3$	$5\bar{1}3$	533
$9\bar{6}6$	$\bar{5}\bar{4}13$	$6\bar{3}3$	$1/2[1\bar{1}2]$	$3\bar{2}2$	$\bar{5}\bar{4}13$	$5\bar{4}3$	$5\bar{3}3$
$15,\bar{6},0$	$7\bar{2}\bar{5}3$	12,3,3	$1/2[712]$	520	$7\bar{2}\bar{5}3$	$7\bar{2}3$	733

(4, 4) = 4 axis, 4 index system; (3, 3) = 3 axis, 3 index system;
(O-H) = orthohexagonal system.

The trigonal unit cell is shown in Figure 5.26. Equations relating the direction indices and the plane indices for rhombohedral indices:

$$\frac{u + (v + w)\cos\alpha}{h} = \frac{v + (w + u)\cos\alpha}{k} = \frac{w + (u + v)\cos\alpha}{1}, \qquad (5.129)$$

and

$$\frac{u}{h + (h - k - l)\cos\alpha} = \frac{v}{k + (k - l - h)\cos\alpha} = \frac{w}{1 + (l - h - k)\cos\alpha}. \qquad (5.130)$$

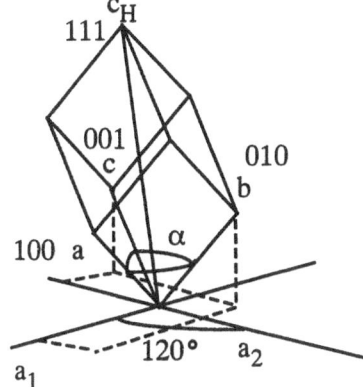

Figure 5.26. Trigonal unit cell showing the principal directions.

5.8. Hexagonal System

The hexagonal system poses several problems for definition of axes, indices, unit vectors, and the related equations describing the structure. There are two definitions of axes widely in use and a third that has some specific advantages. These definitions are:

Three axis, three index	(3, 3) (Miller)
Four axis, four index	(4, 4) (Miller-Bravais)
Orthohexagonal	(O-H)

Usage varies mostly between the Miller and the Miller-Bravais notations. In addition, there is a contracted notation using a dot in place of the i for plane indices or the t for direction indices. This contracted notation leads to misinterpretation of the direction indices, since these differ in the Miller and the Miller-Bravais notations. It is best to avoid using the contracted dot notation, unless the use of Miller or Miller-Bravais indices is explicitly stated.

Another source of confusion is the Miller and Miller-Bravais name. For the occasional user or the student starting to learn the hexagonal or the trigonal systems, there is a tendency to reverse the notation. To avoid this the Miller indices will be referred to as (3, 3) and the Miller-Bravais indices as (4, 4) to clearly indicate the number of axes in the basis vector set. When calculations are to be done or extinctions found, it is best to use the (3, 3) indices and then convert to (4, 4). There is a little more work involved, but the savings in time resulting from fewer errors will more than offset this initial effort.

To minimize confusion among these definitions, the hcp system equations are presented in parallel as far as possible. See Figures 5.27 through 5.31 for coordinate geometries for direct and reciprocal spaces.

Definition of axes:

$$(3, 3): \quad a_1 = a\frac{\sqrt{3}}{2}i + \frac{a}{2}j, \qquad a_2 = -a\frac{\sqrt{3}}{2}i + \frac{a}{2}j, \qquad a_3 = ck, \qquad (5.131)$$

$$(4, 4):$$

$$a_1 = \frac{a}{2}i - a\frac{\sqrt{3}}{2}j, \quad a_2 = \frac{a}{2}i + a\frac{\sqrt{3}}{2}j, \quad a_3 = -(a_1 + a_2) = -ai, \quad a_4 = ck,$$
$$(5.132)$$

$$(O\text{-}H): \quad a_1 = a\frac{\sqrt{3}}{2}i + \frac{a}{2}j, \qquad a_2 = -a\frac{\sqrt{3}}{2}i + \frac{3a}{2}j, \qquad a_3 = ck. \qquad (5.133)$$

Various dot products:

$$(3, 3): \quad a_1 \cdot a_1 = a^2, \qquad a_2 \cdot a_1 = -\frac{a^2}{2}, \qquad a_3 \cdot a_1 = 0,$$

$$a_1 \cdot a_2 = -\frac{a^2}{2}, \qquad a_2 \cdot a_2 = a^2, \qquad a_3 \cdot a_2 = 0,$$

$$a_1 \cdot a_3 = 0, \qquad a_2 \cdot a_3 = 0, \qquad a_3 \cdot a_3 = c^2, \qquad (5.134)$$

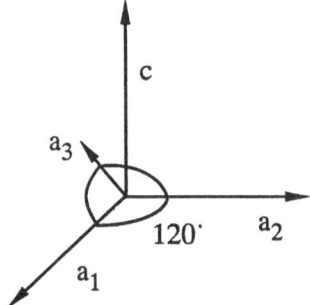

Figure 5.27. Coordinates for (3, 3) are defined by a_1 and a_2; a_3 is not used. Coordinates for (4, 4) are defined by a_1, a_2, a_3 and $a_3 = -a_1 - a_2$. In both coordinate systems c is normal to the other axes.

$(4, 4)$:

$$a_1 \cdot a_1 = a^2, \qquad a_2 \cdot a_1 = -\frac{a^2}{2}, \qquad a_3 \cdot a_1 = -\frac{a^2}{2}, \qquad a_4 \cdot a_1 = 0,$$

$$a_1 \cdot a_2 = -\frac{a^2}{2}, \quad a_2 \cdot a_2 = a^2, \qquad a_3 \cdot a_2 = -\frac{a^2}{2}, \quad a_4 \cdot a_2 = 0,$$

$$a_1 \cdot a_3 = -\frac{a^2}{2}, \quad a_2 \cdot a_3 = -\frac{a^2}{2}, \quad a_3 \cdot a_3 = a^2, \qquad a_4 \cdot a_3 = 0,$$

$$a_1 \cdot a_4 = 0, \qquad a_2 \cdot a_4 = 0, \qquad a_3 \cdot a_4 = 0, \qquad a_4 \cdot a_4 = c^2,$$

$$(5.135)$$

$(O\text{-}H)$:

$$a_1 \cdot a_1 = a^2, \qquad a_2 \cdot a_1 = 0, \qquad a_3 \cdot a_1 = 0,$$

$$a_1 \cdot a_2 = 0, \qquad a_2 \cdot a_2 = 3a^2, \qquad a_3 \cdot a_2 = 0,$$

$$a_1 \cdot a_3 = 0, \qquad a_2 \cdot a_3 = 0, \qquad a_3 \cdot a_3 = c^2. \qquad (5.136)$$

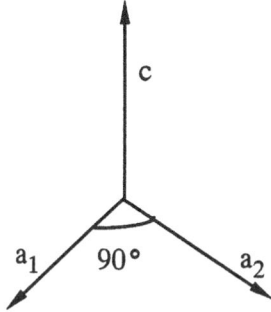

Figure 5.28. Coordinates for (O-H).

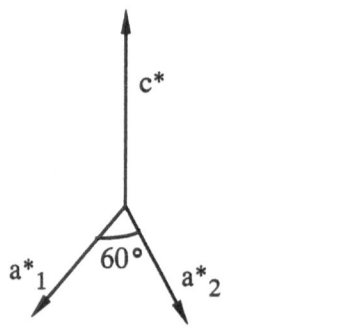

Figure 5.29. Reciprocal lattice coordinates for (3, 3) system.

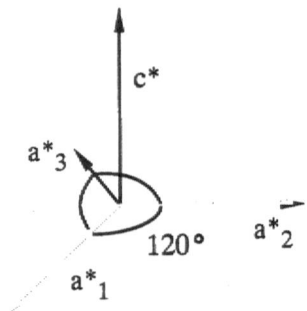

Figure 5.30. Reciprocal lattice coordinates for (4, 4) system.

Reciprocal lattice vector definitions:

(3, 3) :
$$a^*_1 = \frac{2}{3\,a^2}(2\,a_1 + a_2) = \frac{i}{a\sqrt{3}} + \frac{j}{a},$$
$$a^*_2 = \frac{2}{3\,a^2}(a_1 + 2\,a_2) = -\frac{i}{a\sqrt{3}} + \frac{j}{a},$$
$$a^*_3 = \frac{k}{c} = \frac{a_3}{c^2}.$$

(5.137)

(4, 4) :
$$a^*_1 = \frac{i}{3\,a} - \frac{j}{a\sqrt{3}} = \frac{2}{3\,a^2}\,a_1,$$
$$a^*_2 = \frac{i}{3\,a} + \frac{j}{a\sqrt{3}} = \frac{2}{3\,a^2}\,a_2,$$
$$a^*_3 = \frac{2}{3}\frac{i}{a} = \frac{2}{3\,a^2}\,a_3,$$
$$a^*_4 = \frac{k}{c} = \frac{a_4}{c^2}.$$

(5.138)

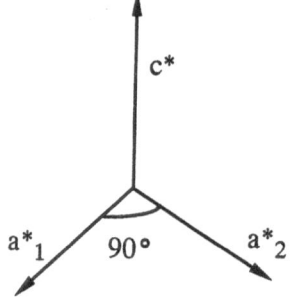

Figure 5.31. Reciprocal lattice coordinates for the (O-H) system.

$$\text{(O-H)}: \quad \mathbf{a^*}_1 = \frac{\sqrt{3}\,\mathbf{i}}{2\,a} + \frac{\mathbf{j}}{2\,a} = \frac{1}{a^2}\,\mathbf{a}_1,$$

$$\mathbf{a^*}_2 = -\frac{\mathbf{i}}{2\sqrt{3}\,a} + \frac{\mathbf{j}}{2\,a} = \frac{1}{3\,a^2}\,\mathbf{a}_2,$$

$$\mathbf{a^*}_3 = \frac{\mathbf{k}}{c} = \frac{\mathbf{a}_3}{c^2}. \tag{5.139}$$

Note that the $\mathbf{a}_i$ differ among the three systems.
   Volume:

(3, 3) $$V = \frac{a^2 c \sqrt{3}}{2}, \tag{5.140}$$

(4, 4): $$V = \frac{a^2 c \sqrt{3}}{2}, \tag{5.141}$$

(O-H): $$V = a^2 c \sqrt{3}. \tag{5.142}$$

   Various dot products:

(3, 3): $$\mathbf{a^*}_1 \cdot \mathbf{a^*}_1 = \frac{4}{3\,a^2}, \quad \mathbf{a^*}_2 \cdot \mathbf{a^*}_1 = \frac{2}{3\,a^2}, \quad \mathbf{a^*}_3 \cdot \mathbf{a^*}_1 = 0,$$

$$\mathbf{a^*}_1 \cdot \mathbf{a^*}_2 = \frac{2}{3\,a^2}, \quad \mathbf{a^*}_2 \cdot \mathbf{a^*}_2 = \frac{4}{3\,a^2}, \quad \mathbf{a^*}_3 \cdot \mathbf{a^*}_2 = 0,$$

$$\mathbf{a^*}_1 \cdot \mathbf{a^*}_3 = 0, \quad\quad \mathbf{a^*}_2 \cdot \mathbf{a^*}_3 = 0, \quad\quad \mathbf{a^*}_3 \cdot \mathbf{a^*}_3 = \frac{1}{c^2}, \tag{5.143}$$

(4, 4):

$$\mathbf{a^*}_1 \cdot \mathbf{a^*}_1 = \frac{4}{9\,a^2}, \quad \mathbf{a^*}_2 \cdot \mathbf{a^*}_1 = -\frac{2}{9\,a^2}, \quad \mathbf{a^*}_3 \cdot \mathbf{a^*}_1 = -\frac{2}{9\,a^2}, \quad \mathbf{a^*}_4 \cdot \mathbf{a^*}_1 = 0,$$

$$\mathbf{a^*}_1 \cdot \mathbf{a^*}_2 = -\frac{2}{9\,a^2}, \quad \mathbf{a^*}_2 \cdot \mathbf{a^*}_2 = \frac{4}{9\,a^2}, \quad \mathbf{a^*}_3 \cdot \mathbf{a^*}_2 = -\frac{2}{9\,a^2}, \quad \mathbf{a^*}_4 \cdot \mathbf{a^*}_2 = 0,$$

$$\mathbf{a^*}_1 \cdot \mathbf{a^*}_3 = -\frac{2}{9\,a^2}, \quad \mathbf{a^*}_2 \cdot \mathbf{a^*}_3 = -\frac{2}{9\,a^2}, \quad \mathbf{a^*}_3 \cdot \mathbf{a^*}_3 = \frac{4}{9\,a^2}, \quad \mathbf{a^*}_4 \cdot \mathbf{a^*}_3 = 0,$$

$$\mathbf{a^*}_1 \cdot \mathbf{a^*}_4 = 0, \quad\quad \mathbf{a^*}_2 \cdot \mathbf{a^*}_4 = 0, \quad\quad \mathbf{a^*}_3 \cdot \mathbf{a^*}_4 = 0, \quad\quad \mathbf{a^*}_4 \cdot \mathbf{a^*}_4 = \frac{1}{c^2},$$

$$\tag{5.144}$$

(O-H): $$\mathbf{a^*}_1 \cdot \mathbf{a^*}_1 = \frac{1}{a^2}, \quad\quad \mathbf{a^*}_2 \cdot \mathbf{a^*}_1 = 0, \quad\quad \mathbf{a^*}_3 \cdot \mathbf{a^*}_1 = 0,$$

$$\mathbf{a^*}_1 \cdot \mathbf{a^*}_2 = 0, \quad\quad \mathbf{a^*}_2 \cdot \mathbf{a^*}_2 = \frac{1}{3\,a^2}, \quad \mathbf{a^*}_3 \cdot \mathbf{a^*}_2 = 0,$$

$$\mathbf{a^*}_1 \cdot \mathbf{a^*}_3 = 0, \quad\quad \mathbf{a^*}_2 \cdot \mathbf{a^*}_3 = 0, \quad\quad \mathbf{a^*}_3 \cdot \mathbf{a^*}_3 = \frac{1}{c^2}. \tag{5.145}$$

Mixed dot products:

(3, 3): $\mathbf{a^*_1 \cdot a_1} = 1,$ $\mathbf{a^*_2 \cdot a_1} = 0,$ $\mathbf{a^*_3 \cdot a_1} = 0,$

$\mathbf{a^*_1 \cdot a_2} = 0,$ $\mathbf{a^*_2 \cdot a_2} = 1,$ $\mathbf{a^*_3 \cdot a_2} = 0,$

$\mathbf{a^*_1 \cdot a_3} = 0,$ $\mathbf{a^*_2 \cdot a_3} = 0,$ $\mathbf{a^*_3 \cdot a_3} = 1,$ (5.146)

(4, 4): $\mathbf{a^*_1 \cdot a_1} = \dfrac{2}{3},$ $\mathbf{a^*_2 \cdot a_1} = -\dfrac{1}{3},$ $\mathbf{a^*_3 \cdot a_1} = -\dfrac{1}{3},$ $\mathbf{a^*_4 \cdot a_1} = 0,$

$\mathbf{a^*_1 \cdot a_2} = -\dfrac{1}{3},$ $\mathbf{a^*_2 \cdot a_2} = \dfrac{2}{3},$ $\mathbf{a^*_3 \cdot a_2} = -\dfrac{1}{3},$ $\mathbf{a^*_4 \cdot a_2} = 0,$

$\mathbf{a^*_1 \cdot a_3} = -\dfrac{1}{3},$ $\mathbf{a^*_2 \cdot a_3} = -\dfrac{1}{3},$ $\mathbf{a^*_3 \cdot a_3} = \dfrac{2}{3},$ $\mathbf{a^*_4 \cdot a_3} = 0,$

$\mathbf{a^*_1 \cdot a_4} = 0,$ $\mathbf{a^*_2 \cdot a_4} = 0,$ $\mathbf{a^*_3 \cdot a_4} = 0,$ $\mathbf{a^*_4 \cdot a_4} = 1,$

(5.147)

(O-H): $\mathbf{a^*_1 \cdot a_1} = 1,$ $\mathbf{a^*_2 \cdot a_1} = 0,$ $\mathbf{a^*_3 \cdot a_1} = 0,$

$\mathbf{a^*_1 \cdot a_2} = 0,$ $\mathbf{a^*_2 \cdot a_2} = 1,$ $\mathbf{a^*_3 \cdot a_2} = 0,$

$\mathbf{a^*_1 \cdot a_3} = 0,$ $\mathbf{a^*_2 \cdot a_3} = 0,$ $\mathbf{a^*_3 \cdot a_3} = 1.$ (5.148)

Some cross products in (3, 3) useful for calculations:
Direct lattice:

$$\mathbf{a_1 \times a_2} = \frac{a^2 \sqrt{3}}{2\,c}\,\mathbf{a_3} = \frac{V\,\mathbf{a_3}}{c^2},$$ (5.149)

$$\mathbf{a_2 \times a_3} = \frac{c}{\sqrt{3}}\,(2\,\mathbf{a_1} + \mathbf{a_2}),$$ (5.150)

$$\mathbf{a_3 \times a_1} = \frac{c}{\sqrt{3}}\,(2\,\mathbf{a_2} + \mathbf{a_1}).$$ (5.151)

Reciprocal lattice:

$$\mathbf{a^*_1 \times a^*_2} = \frac{2\,c}{a^2 \sqrt{3}}\,\mathbf{a^*_3} = \frac{\mathbf{a_3}}{V},$$ (5.152)

$$\mathbf{a^*_2 \times a^*_3} = \frac{1}{c\sqrt{3}}\,(2\,\mathbf{a^*_1} - \mathbf{a^*_2}) = \frac{\mathbf{a_1}}{V},$$ (5.153)

$$\mathbf{a^*_3 \times a^*_1} = \frac{1}{c\sqrt{3}}\,(-\mathbf{a^*_1} + 2\,\mathbf{a^*_2}) = \frac{\mathbf{a_2}}{V}.$$ (5.154)

For $\mathbf{g_1}, \mathbf{g_2}$ any two reciprocal lattice vectors,

$$\mathbf{g_1 \times g_2} = \mathbf{g_3} = \frac{a^2}{2\,V}\,(h_3\,\mathbf{a^*_1} + k_3\,\mathbf{a^*_2}) + \frac{c^2}{V}\,l_3\,\mathbf{a^*_3} = (h_3\ k_3\ l_3),$$ (5.155)

$$h_3 \equiv l_2\,(2\,k_1 + h_1) - l_1\,(2\,k_2 + h_2),$$ (5.156)

$$k_3 \equiv l_1 \, (2 \, h_2 + k_2) - l_2 \, (2 \, h_1 + k_1),$$ (5.157)

$$l_3 \equiv 3 \, (h_1 \, k_2 - k_1 \, h_2).$$ (5.158)

Also note

$$\mathbf{g}_1 \times \mathbf{g}_2 = \mathbf{u}_3 = \frac{1}{V} \, (u_3 \, \mathbf{a}_1 + v_3 \, \mathbf{a}_2 + w_3 \, \mathbf{a}_3),$$ (5.159)

$$u_3 = k_1 \, l_2 - l_1 \, k_2,$$ (5.160)

$$v_3 = l_1 \, h_2 - h_1 \, l_2,$$ (5.161)

$$w_3 = h_1 \, k_2 - k_1 \, h_2.$$ (5.162)

Indices of planes in the three indexing systems are

$(3, 3)$: $H \, a^*_1 + K \, a^*_2 + L \, a^*_3,$ (5.163)

$(4, 4)$: $h \, a^*_1 + k \, a^*_2 + i \, a^*_3 + l \, a^*_4,$ (5.164)
$\quad\quad\quad\quad i = - h - k,$ (5.164a)

$(O\text{-}H)$: $H^o \, a^*_1 + K^o \, a^*_2 + L^o \, a^*_3.$ (5.165)

Indices of directions in the three indexing systems are:

$(3, 3)$: $U \, \mathbf{a}_1 + V \, \mathbf{a}_2 + W \, \mathbf{a}_3,$ (5.166)

$(4, 4)$: $u \, \mathbf{a}_1 + v \, \mathbf{a}_2 + t \, \mathbf{a}_3 + w \, \mathbf{a}_4,$ (5.167)
$\quad\quad\quad\quad t = - u - v,$ (5.167a)

$(O\text{-}H)$: $U^o \, \mathbf{a}_1 + V^o \, \mathbf{a}_2 + W^o \, \mathbf{a}_3.$ (5.168)

Relationships among the indices:
Given $(3, 3)$ indices U, V, W. To find $(4, 4)$ indices u, v, w:

$$u = \frac{1}{3} \, (2 \, U - V), \quad v = \frac{1}{3} \, (2 \, V - U), \quad w = W.$$ (5.169)

For plane indices (HKL) and (hkl) :

$$h = H, \quad\quad\quad k = K, \quad\quad\quad l = L.$$ (5.170)

Given $(4, 4)$ indices u, v, t, w. To find $(3, 3)$ indices U, V, W:

$$U = 2 \, u + v, \quad\quad V = 2 \, v + u, \quad\quad W = w.$$ (5.171)

To convert to or from $(O\text{-}H)$:

$$U = U^o + V^o, \quad\quad V = 2 \, V^o, \quad\quad W = W^o,$$ (5.172)

$$U^o = U - \frac{V}{2}, \qquad V^o = \frac{V}{2}, \qquad W^o = W; \tag{5.173}$$

$$U^o = \frac{3u}{2}, \qquad V^o = \frac{u}{2} + v, \qquad W^o = w; \tag{5.174}$$

$$u = \frac{2}{3}U^o, \qquad v = V^o - \frac{U^o}{2}, \quad w = W^o; \tag{5.175}$$

$$H = H^o, \qquad K = \frac{1}{2}(K^o - H^o), \qquad L = L^o; \tag{5.176}$$

$$H^o = H, \qquad K^o = 2K + H, \qquad L^o = L; \tag{5.177}$$

$$H^o = h, \qquad K^o = 2k + h, \qquad L^o = l; \tag{5.178}$$

$$h = H^o, \qquad k = \frac{K^o - H^o}{2}, \qquad l = L^o. \tag{5.179}$$

Relationships between direction indices and plane indices for the three indexing systems (in these equations γ is the c/a ratio):
For the (3, 3) system:

$$\frac{U}{2H + K} = \frac{V}{H + 2K} = \frac{W}{\left(\frac{3}{2}\gamma^{-2}\right)L}, \tag{5.180}$$

$$\frac{2U - V}{H} = \frac{2V - U}{K} = \frac{(2\gamma^2)W}{L}. \tag{5.181}$$

For the (4, 4) system:

$$\frac{u}{h} = \frac{v}{k} = \frac{w}{\left(\frac{3}{2}\gamma^{-2}\right)l}. \tag{5.182}$$

For the (O-H) system:

$$\frac{U^o}{H^o} = \frac{3V^o}{K^o} = \frac{W^o\gamma^2}{L^o}. \tag{5.183}$$

The distance, d (in Å or nm), between equivalent planes for the three systems is

$$(3, 3): \qquad \frac{a}{\sqrt{\frac{4}{3}(H^2 + K^2 + HK) + \gamma^{-2}L^2}}, \tag{5.184}$$

$$(4, 4): \qquad \frac{a}{\sqrt{\frac{4}{3}(h^2 + k^2 + hk) + \gamma^{-2}l^2}}, \tag{5.185}$$

(O-H) :
$$\frac{a}{\sqrt{\frac{4}{3}(H^{o2} + \frac{1}{3}K^{o2}) + \gamma^{-2}L^{o2}}}.$$
(5.186)

The magnitude of the reciprocal lattice vector, **g**, for the plane is $1/d$ (the units are Å^{-1} or nm^{-1}).

The distance between equivalent planes in the direct lattice, i.e., the repeat distance $|\mathbf{u}|$, is

(3, 3) : $\qquad a\sqrt{U^2 + V^2 - UV + \gamma^2 W^2},$ (5.187)

(4, 4) : $\qquad a\sqrt{3(u^2 + v^2 + uv) + \gamma^2 w^2},$ (5.188)

(O-H) : $\qquad a\sqrt{U^{o2} + 3V^{o2} + \gamma^2 W^{o2}}.$ (5.189)

The repeat distance in reciprocal space is defined as $H = 1/|\mathbf{u}|$.

Angle between two planes for each system:

(3, 3) : $\quad \cos\theta_{12} = \frac{4}{3a^2}\left[H_1H_2 + K_1K_2 + \frac{1}{2}(H_1K_2 + H_2K_1) + \frac{3}{4}\gamma^{-2}L_1L_2\right]d_1d_2,$

(5.190)

(4, 4) : $\quad \cos\theta_{12} = \frac{4}{3a^2}\left[h_1h_2 + k_1k_2 + \frac{1}{2}(h_1k_2 + h_2k_1) + \frac{3}{4}\gamma^{-2}l_1l_2\right]d_1d_2,$ (5.191)

(O-H) : $\quad \cos\theta_{12} = \frac{1}{a^2}\left(H^o_1 H^o_2 + \frac{1}{3}K^o_1 K^o_2 + \gamma^{-2}L^o_1 L^o_2\right)d_1 d_2,$ (5.192)

where d_1 and d_2 are the repeat distances for plane indices in the appropriate system.

The angle between directions is

(3, 3) :

$$\cos\theta_{12} = \frac{a^2\left[U_1U_2 + V_1V_2 - \frac{1}{2}(U_1V_2 + U_2V_1) + \gamma^2 W_1W_2\right]}{n_1n_2},$$ (5.193)

(4, 4) :

$$\cos\theta_{12} = \frac{3a^2\left[u_1u_2 + v_1v_2 + \frac{1}{2}(u_1v_2 + u_2v_1) + \frac{\gamma^2}{3}w_1w_2\right]}{n_1n_2},$$ (5.194)

(O-H) :

$$\cos\theta_{12} = a^2(U^o_1 U^o_2 + 3V^o_1 V^o_2 + \gamma^2 W^o_1 W^o_2)n_1^{-1}n_2^{-1},$$ (5.195)

where n_1 and n_2 are the repeat distances, $|\mathbf{u}|$, for direction indices in the appropriate system.

Indices for various hexagonal indexing systems are listed in Table 5.2. Direction indices are listed in the first three columns, and plane indices are listed in the last three columns. Factors of three have been left in the indices to emphasize the fact that the vectors are not equal but only equivalent, because the magnitudes are not equal, although the direction of each vector is the same.

Projections of (0001) hcp and (001) O-H are shown in Figures 5.32 and 5.33.

Table 5.2. Indices for Various Hexagonal Indexing Systems.

Direction Indices [ ]			Plane Indices ( )		
(4, 4)	(3, 3)	(O-H)	(4, 4)	(3, 3)	(O-H)
[0001]	[001]	[001]	(0001)	(001)	(001)
$11\bar{2}0$	330	330	$11\bar{2}0$	110	130
$\bar{1}2\bar{1}0$	030	$1/2[\bar{3}30]$	$\bar{1}2\bar{1}0$	$\bar{1}20$	$\bar{1}30$
$2\bar{1}\bar{1}0$	300	300	$2\bar{1}\bar{1}0$	$2\bar{1}0$	200
$\bar{1}2\bar{1}1$	031	$1/2[\bar{3}32]$	$\bar{1}2\bar{1}1$	$\bar{1}21$	$\bar{1}31$
$\bar{1}2\bar{1}6$	036	$1/2[\bar{1}\bar{1}4]$	$\bar{1}2\bar{1}6$	$\bar{1}26$	$\bar{1}36$
$11\bar{2}3$	333	$1/2[336]$	$11\bar{2}3$	113	133
$\bar{1}2\bar{1}3$	033	$1/2[\bar{3}36]$	$\bar{1}2\bar{1}3$	$\bar{1}23$	$\bar{1}33$
$\bar{2}113$	303	303	$\bar{2}113$	$\bar{2}13$	$\bar{2}03$
$\bar{2}4\bar{2}3$	063	$\bar{1}11$	$\bar{2}113$	$.\bar{2}13$	$\bar{2}03$
$10\bar{1}0$	210	$1/2[310]$	$10\bar{1}0$	100	110
$01\bar{1}0$	120	010	$01\bar{1}0$	010	020
$\bar{1}100$	$\bar{1}10$	$1/2[\bar{3}10]$	$\bar{1}100$	$\bar{1}10$	$\bar{1}10$
$10\bar{1}1$	211	$1/2[312]$	$10\bar{1}1$	101	111
$01\bar{1}\bar{1}$	$12\bar{1}$	$01\bar{1}$	$01\bar{1}1$	$01\bar{1}$	$02\bar{1}$
$\bar{1}101$	$\bar{1}11$	$1/2[\bar{3}12]$	$\bar{1}101$	$\bar{1}11$	$\bar{1}11$
$20\bar{2}3$	423	313	$20\bar{2}3$	203	223
$02\bar{2}3$	243	023	$02\bar{2}3$	023	043
$\bar{2}203$	$\bar{2}23$	$\bar{3}13$	$\bar{2}203$	$\bar{2}23$	$\bar{2}23$
$10\bar{1}2$	212	$1/2[314]$	$10\bar{1}2$	102	112
$01\bar{1}2$	122	012	$01\bar{1}2$	012	022
$1\bar{1}02$	$1\bar{1}2$	$1/2[3\bar{1}4]$	$1\bar{1}02$	$1\bar{1}2$	$1\bar{1}2$
$12\bar{3}0$	450	$1/2[350]$	$12\bar{3}0$	120	150
$5\bar{1}\bar{4}0$	930	$1/2[510]$	$5\bar{1}\bar{4}0$	$5\bar{1}0$	530
$5\bar{4}\bar{1}0$	$6\bar{3}0$	$1/2[5\bar{1}0]$	$5\bar{4}\bar{1}0$	$5\bar{4}0$	$5\bar{3}0$
$5\bar{1}\bar{4}3$	933	$1/2[512]$	$5\bar{1}\bar{4}3$	$5\bar{1}3$	533
$5\bar{4}\bar{1}3$	$6\bar{3}3$	$1/2[1\bar{1}2]$	$5\bar{4}\bar{1}3$	$5\bar{4}3$	$5\bar{3}3$
$7\bar{2}\bar{5}3$	12,3,3	$1/2[712]$	$7\bar{2}\bar{5}3$	$7\bar{2}3$	733

(4, 4) = 4 axis, 4 index system; (3, 3) = 3 axis, 3 index system;
(O-H) = orthohexagonal system.

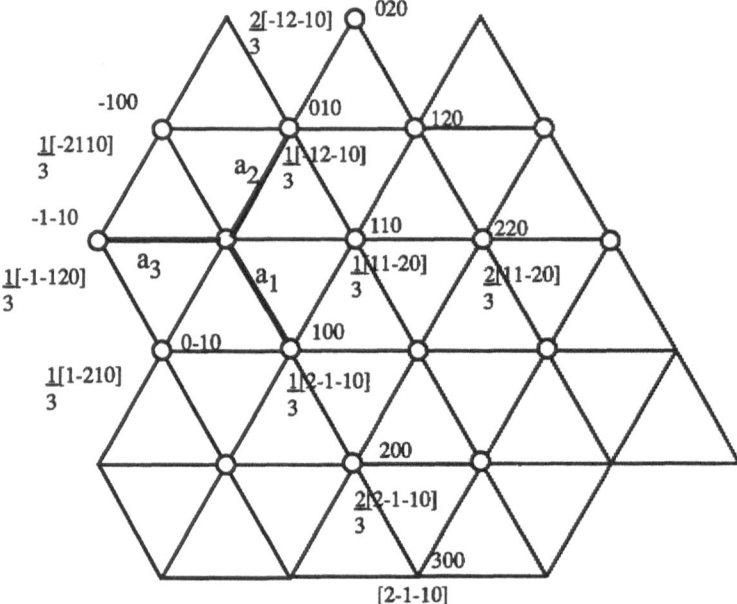

Figure 5.32. Basal plane projection of the hcp lattice indexed for (3, 3) and for (4, 4) axes. Note the fractional vectors in the (4, 4) system which are equal to full vectors in the (3, 3) system. This 3-factor is usually factored out, but it must be considered when calculations are done.

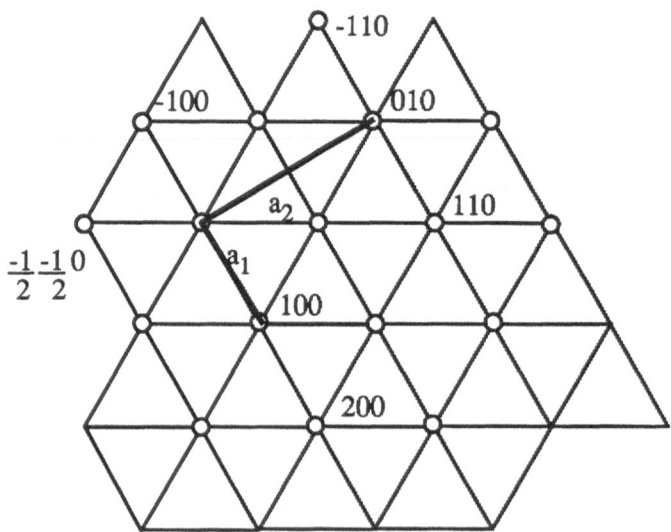

Figure 5.33. Projection along the [001] axis of the hcp system expressed in the orthohexagonal basis vector coordinates. Note that $|\mathbf{a}_2| = \sqrt{3}\,|\mathbf{a}_1|$.

6
Conversion Formulas

6.1. Introduction

Representation of a lattice in its primitive form is necessary to identify conditions under which extinctions may occur. Transformation to another lattice centering or a different lattice system may also be advantageous to understanding interactions to which the crystal may have been subjected.

The general method for accomplishing such transformations is described fully in the International Crystallography Tables [ITC, 1983, p. 70-79]. The approach is to use transformation matrices, and a table of possible transformations is included as part of the ITC description. The transformations included in the following sections are specific examples of the general transformations. These examples are of interest because of special conditions that may prevail, allowing unique transformations to be made. Some transformations in ordered lattices are presented in Chapter 9.

Although special relationships among the lattice parameters are not required (see ITC), conversions that allow lattices to be interchanged with another can be made from BCC to hcp, hcp to orthorhombic, and BCC to orthorhombic if special relationships among the lattice parameters hold, as listed in Table 6.1. Transformations from the primitive cubic to FCC or BCC are covered in detail in Chapter 2. See Chapter 5 for additional details of transformations between trigonal or rhombohedral reference frames and hcp.

6.2. BCC to Orthorhombic

The transformation from BCC to an orthorhombic unit cell is accomplished by applying the transformation defined through the following set of vector equations, where the BCC basis vector set is that defined in Chapter 5. The magnitude of the BCC vector is a_B, the lattice parameter of the BCC unit cell. The conversion equations are:

$$a^o = a_{1B} + a_{2B}, \quad b^o = -a_{1B} + a_{2B} + 2a_{3B}, \quad c^o = \frac{1}{2}a_{1B} - \frac{1}{2}a_{2B} + \frac{1}{2}a_{3B}. \qquad (6.1)$$

The various dot products among the orthorhombic vectors in terms of the BCC lattice

Table 6.1. S·ecial Transformations for BCC, hcp, and Orthorhombic Lattices.

BCC to hcp	hcp to Orthorhombic	BCC to Orthorhombic
$a_H = \sqrt{2}\, a_B$	$a^o = a_H$	$a^o = a_B$
$c_H = a_B \sqrt{3}\,/\,2$	$b^o = a_H\sqrt{3}$	$b^o = a_B \sqrt{2}$
	$c^o = c_H$	$c^o = a_B \sqrt{2}$

a^o, b^o, c^o refer to orthorhombic; a_H, c_H refer to hcp; a_B refers to BCC.

parameter are given by the following set of relationships:

$$a^o \cdot a^o = 2\, a_B, \quad b^o \cdot a^o = 0, \quad\quad c^o \cdot a^o = 0,$$

$$a^o \cdot b^o = 0, \quad\quad b^o \cdot b^o = 6\, a_B, \quad c^o \cdot b^o = 0,$$

$$a^o \cdot c^o = 0, \quad\quad b^o \cdot c^o = 0, \quad\quad c^o \cdot c^o = \frac{3}{4}\, a_B. \tag{6.2}$$

Note that these equations are also used for the BCC to orthorhombic transformation via the BCC to hcp transformation, because

$$a^o = a_W = \sqrt{2}\, a_B, \quad b^o = \sqrt{3}\, a_W = \sqrt{3}\sqrt{2}\, a_B, \quad c^o = a_W = \frac{\sqrt{3}}{2}\, a_B, \tag{6.3}$$

where a_W is the "a" parameter in hcp.

6.3. hcp to Orthorhombic

Define a set of hexagonal vectors in terms of the vectors of an orthogonal coordinate system by

$$a_{1B} = \frac{1}{\sqrt{3}}(2\,a_1 - a_2 - a_3),$$

$$a_{2B} = \frac{1}{\sqrt{3}}(-a_1 + 2\,a_2 - a_3),$$

$$c_B = \frac{1}{2}(a_1 + a_2 + a_3),$$

$$|a_B| = |a_{1B}| = |a_{2B}| = \sqrt{2}\, a, \quad |c_B| = \frac{\sqrt{3}}{2}\, a, \tag{6.4}$$

where a is the repeat parameter. From these relations, we obtain

$$a_{1H} \cdot a_{1H} = 2\, a^2, \quad a_{2H} \cdot a_{1H} = -a^2, \quad c_H \cdot a_{1H} = 0,$$

$$a_{1H} \cdot a_{2H} = -a^2, \quad a_{2H} \cdot a_{2H} = 2\, a^2, \quad c_H \cdot a_{2H} = 0,$$

$$a_{1H} \cdot c_H = 0, \quad a_{2H} \cdot c_H = 0, \quad c_H \cdot c_H = \frac{3}{4}\, a^2. \tag{6.5}$$

Use these hexagonal vectors to define a set of orthorhombic basis vectors:

$$a^o = a_{1B}, \quad b_o = a_{1B} + 2\, a_{2B}, \quad c_o = c_B. \tag{6.6}$$

Then the magnitudes of the vectors are

$$|a^o| = |a_{1B}|, \quad |b^o| = \sqrt{3}\,|a_{1B}|, \quad |c^o| = |c_B|. \tag{6.7}$$

Hence, an orthorhombic system with axes defined in terms of a reference hcp system has lattice parameters

$$a^o = a_H, \quad b^o = \sqrt{3}\, a_H, \quad c^o = c_H. \tag{6.8}$$

If the hcp has been produced by a BCC to hcp transformation (see Section 6.4), then

$$a^o = a_H = \sqrt{2}\, a_B, \quad b^o = \sqrt{3}\, a_H = \sqrt{6}\, a_B, \quad c^o = c_H = \frac{\sqrt{3}}{2} a_B. \tag{6.9}$$

If the reference axes are in the (4, 4) form, then the resulting transformation is not changed. Hence, either (3, 3) or (4, 4) axes can be used to obtain the same results, but the use of the indices, of course, still must follow the rules for the (3, 3) or the (4, 4) notation, as described in Section 5.8.

6.4. BCC to hcp

In this transformation, the hcp axes are derived from vectors along the BCC cube faces and along the body diagonal:

$$a_{1B} = -b + c, \quad a_{2B} = a - c, \quad c_B = \frac{1}{2}(a + b + c), \tag{6.10}$$

and the dot product relationships are

$$a_{1H} \cdot a_{1H} = 2\, a_B^2, \quad a_{2H} \cdot a_{1H} = -a_B^2, \quad c_H \cdot a_{1H} = 0,$$

$$a_{1H} \cdot a_{2H} = -a_B^2, \quad a_{2H} \cdot a_{2H} = 2\, a_B^2, \quad c_H \cdot a_{2H} = 0,$$

$$a_{1H} \cdot c_H = 0, \quad a_{2H} \cdot c_H = 0, \quad c_H \cdot c_H = \frac{3}{4} a_B. \tag{6.11}$$

6.5. FCT to BCT and BCT to FCT Transformation

6.5.1. FCT to BCT

The relevant vectors are

$$a' = a, \qquad b' = \frac{1}{2} b + c \qquad c' = -\frac{1}{2} b + \frac{1}{2} c, \tag{6.12a}$$

where a, b, c are the FCT basis vectors, and

$$|a'| = a, \quad |b'| = \frac{1}{2}\sqrt{a^2 + c^2} = |c'|. \tag{6.12b}$$

The dot products are

$$a' \cdot a' = a^2, \quad b' \cdot a' = 0, \qquad c' \cdot a' = 0,$$

$$a' \cdot b' = 0, \quad b' \cdot b' = \frac{1}{4}(a^2 + c^2), \quad c' \cdot b' = \frac{1}{4}(-a^2 + c^2),$$

$$a' \cdot c' = 0, \quad b' \cdot c' = \frac{1}{4}(-a^2 + c^2), \quad c' \cdot c' = \frac{1}{4}(a^2 + c^2). \tag{6.13}$$

Figure 6.1. Vectors in the transformation of FCT to BCT. (a, b, c) in FCT and (a', b', c') in BCT. Projection is the (100) plane, and a and a' coincide. Circles are in the plane; squares are at z/2 above the plane.

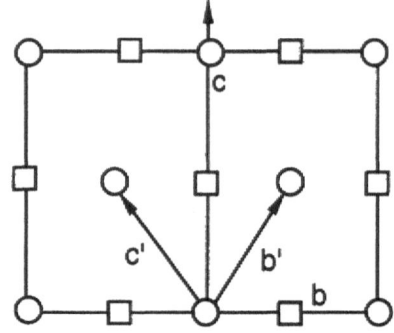

6.5.2. BCT to FCT

The vectors are

$$\mathbf{a} = \mathbf{a'}, \qquad \mathbf{b} = \mathbf{b'} - \mathbf{c'}, \qquad \mathbf{c} = \mathbf{b'} + \mathbf{c'}. \tag{6.14}$$

The dot products are

$$
\begin{array}{lll}
\mathbf{a} \cdot \mathbf{a} = a^2, & \mathbf{b} \cdot \mathbf{a} = 0, & \mathbf{c} \cdot \mathbf{a} = 0, \\
\mathbf{a} \cdot \mathbf{b} = 0, & \mathbf{b} \cdot \mathbf{b} = a^2, & \mathbf{c} \cdot \mathbf{b} = 0, \\
\mathbf{a} \cdot \mathbf{c} = 0, & \mathbf{b} \cdot \mathbf{c} = 0, & \mathbf{c} \cdot \mathbf{c} = c^2.
\end{array}
\tag{6.15}
$$

The transformation is shown in Figure 6.1.

6.6. Monoclinic Nonprimitive to Primitive Transformation

For a B face-centering (b axis unique), the relevant nonprimitive vectors are

$$\mathbf{a'} = \frac{1}{2}\mathbf{a} + \frac{1}{2}\mathbf{c}, \quad \mathbf{b'} = \mathbf{b}, \quad \mathbf{c'} = -\frac{1}{2}\mathbf{a} + \frac{1}{2}\mathbf{c}, \tag{6.16}$$

where $\mathbf{a}, \mathbf{b}, \mathbf{c}$ are the nonprimitive basis vectors (see Figure 6.2), and the basis vector dot products are given by

$$\mathbf{a'} \cdot \mathbf{a'} = \frac{1}{4}(a^2 + c^2 + 2\,a\,c\,\cos\beta), \quad \mathbf{b'} \cdot \mathbf{a'} = 0, \quad \mathbf{c'} \cdot \mathbf{a'} = \frac{1}{4}(c^2 - a^2),$$

$$\mathbf{a'} \cdot \mathbf{b'} = 0, \qquad\qquad\qquad \mathbf{b'} \cdot \mathbf{b'} = b^2, \quad \mathbf{c'} \cdot \mathbf{b'} = 0,$$

$$\mathbf{a'} \cdot \mathbf{c'} = \frac{1}{4}(c^2 - a^2), \qquad\qquad \mathbf{b'} \cdot \mathbf{c'} = 0, \quad \mathbf{c'} \cdot \mathbf{c'} = \frac{1}{4}(a^2 + c^2 - 2\,a\,c\,\cos\beta).$$

$$\tag{6.17}$$

Figure 6.2.   Basis vectors for
nonprimitive monoclinic (**a, b, c**) and
primitive (**a', b', c'**). The a-c angle β
transforms to the a'-c' angle β'.

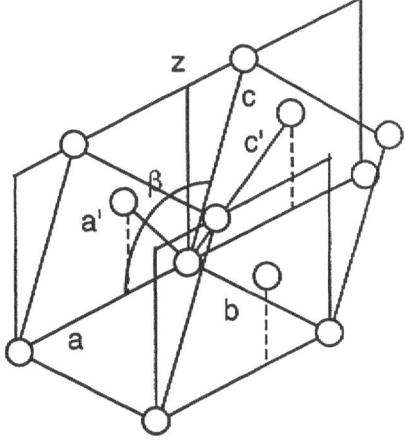

The new angle in terms of the parameters of the nonprimitive basis is given by

$$\cos \beta' = \frac{-a^2 + c^2}{\sqrt{(a^2 + c^2 + 2ac \cos \beta)(a^2 + c^2 - 2ac \cos \beta)}} \ . \tag{6.18}$$

6.7. Rhombohedral to hcp

The transformation from a rhombohedral reference basis to an hcp reference basis
(triple hexagonal cell, obverse setting [ITC, 1983, p. 56]) is accomplished using the
transformation equations

$$a_{1H} = a_{2R} - a_{3R}, \quad a_{2H} = a_{1R} - a_{3R}, \quad a_{3H} = a_{1R} + a_{2R} + a_{3R}, \tag{6.19}$$

and the magnitudes of the hcp vectors in terms of the rhombohedral magnitudes are

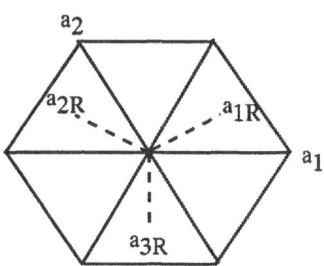

Figure 6.3. Triple hexagonal unit cell and the primitive rhombohedral cell, in (0001)
projection for hcp and (111) projection for rhombohedral. The hcp vectors are a_1, a_2,
and c; the rhombohedral vectors are a_{1R}, a_{2R}, and a_{3R} [ITC, 1983, p. 57]. The
coordinates of the rhombohedral vectors in the hcp lattice are:

$$a_{1R} = \frac{2}{3} \ \frac{1}{3} \ \frac{1}{3} \ ; \quad a_{2R} = -\frac{1}{3} \ \frac{2}{3} \ \frac{1}{3} \ ; \quad a_{3R} = -\frac{1}{3} \ -\frac{2}{3} \ \frac{1}{3} \ .$$

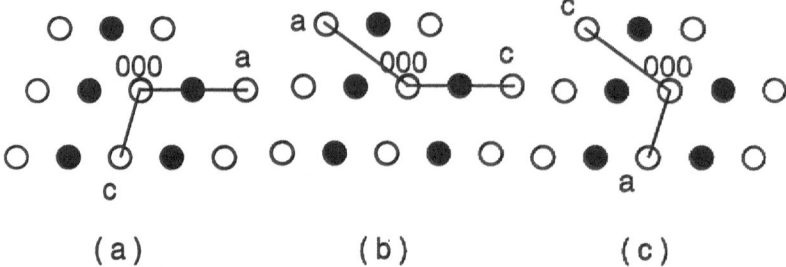

Figure 6.4. Monoclinic basis vectors for various centerings. The projection is (010), b out of the page; white circles lie in the plane, and black circles lie z/2 above the plane.

$$|a_{1H}| = |a_{2H}| = a_H = a_R \sqrt{2(1 - \cos \alpha)}, \tag{6.20}$$

and

$$|a_{3H}| = |c_H| = a_R \sqrt{3(1 + 2\cos \alpha)}. \tag{6.21}$$

In Figure 6.3 is shown the triple hexagonal cell. The monoclinic centerings are shown in Figure 6.4.

6.8. Some Orientation Relationships

The relationships were compiled from Dahmen [1982].

Kurdjumov-Sachs:.......................... (110)BCC ∥ (111)FCC; [Ī11]BCC ∥ [Ī10]FCC

Nishijima-Wasserman: .... (110)BCC ∥ (111)FCC; [001]BCC ∥ [Ī01]FCC
and [Ī10]BCC ∥ [Ī2Ī]FCC

Bain: .... (010)BCC ∥ (010)FCC; [001]BCC ∥ [Ī01]FCC
and [101]BCC ∥ [001]FCC

Pitsch (inverse N-W):..................... (101)BCC ∥ (001)FCC; [Ī11]BCC ∥ [Ī10]FCC

Burgers:... (110)BCC ∥ (0001)hcp; [Ī11]BCC ∥ [Ī2Ī0]hcp
and [Ī1Ī]BCC ∥ [10Ī0]hcp

Pitsch-Schrader:............................. (110)BCC ∥ (0001)hcp; [Ī10]BCC ∥ [01Ī0]hcp

Omega:... (111)BCC ∥ (0001)hcp; [110]BCC ∥ [11Ī0]hcp

6.9. Some Ordered Structures

Transformations from disordered to ordered structures for BCC, FCC, and hcp have been divided into subgroups by Tanner and Leamy [1974]. See Table 6.2.

Isostructural transformations are those transformations in which no change of crystal system occurs. **Neostructural transformations** are the transformations in which the crystal system changes. See Figure 6.5.

Table 6.2. Disordered to Ordered Transformations.

Lattice	Symbol	Lattice parameter	
		disordered	ordered
BCC = A2:	$A2 \Rightarrow B2$	a	a
	$A2 \Rightarrow DO_3$	a	2a
FCC:	$A1 \Rightarrow L1_2$	a	a
HCP:	$A3 \Rightarrow DO_{19}$	a,c	2a,c
FCC=A1:	$A1 \Rightarrow L1_0$	(a,a,a)	(a,a,c)
	$A1 \Rightarrow$ Long Period Superlattice (LPS):		
	(CuAuII)	(a,a,a)	(a,multiple,a)
	$A1 \Rightarrow DO_{22}$	(a,a,a)	(a,a,2a)
	$A1 \Rightarrow D1_a$	(a,a,a)	$\left(\dfrac{\sqrt{10}}{2}a, \dfrac{\sqrt{10}}{2}a, a\right)$
	$A1 = D_{2h}^{25} I_{mmm}$	(a,a,a)	$\left(\dfrac{3\sqrt{2}}{2}a, \dfrac{1}{\sqrt{2}}a, a\right)$
	$A1 \Rightarrow L1_1$	a	2a (rhombohedral, α)

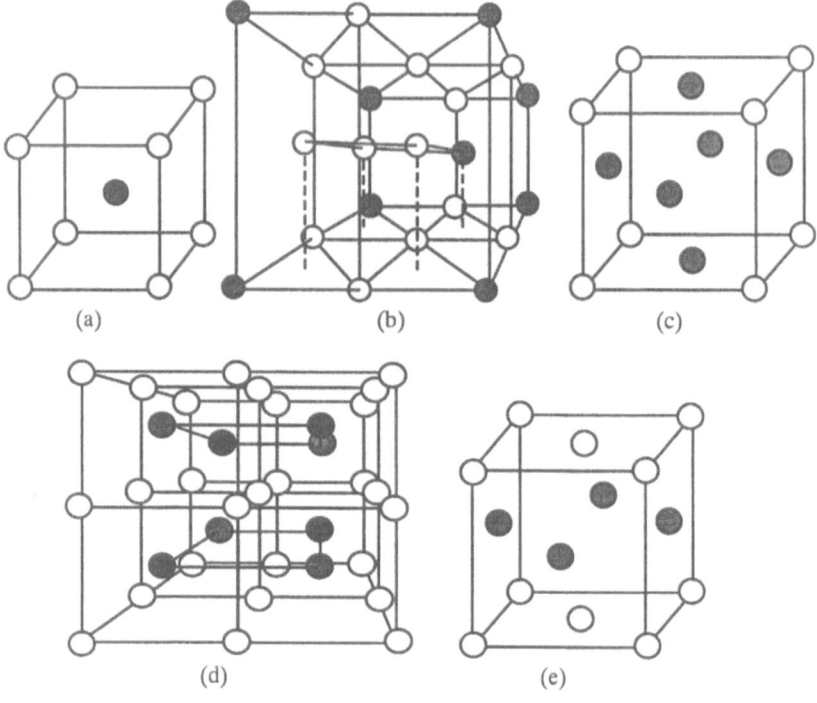

(a)   (b)   (c)

(d)   (e)

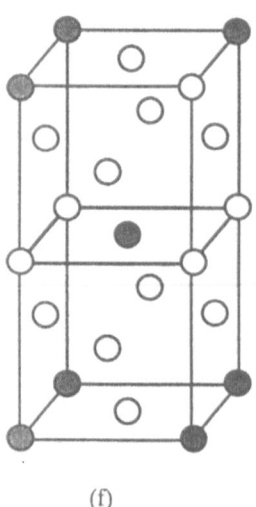

(f)

Figure 6.5.   Examples of isostructural ordering transformations based on the Tanner and Leamy classification of isostructural (a)-(d) and neostructural (e)-(f) transformations. (a) B2 ; (b) DO_{19}; (c) $L1_2$; (d) DO_3 ; (e) $L1_0$; (f) DO_{22} .

7
Slip Systems

7.1. Face-Centered Cubic

Slip occurs on {111} planes (close packed planes) along <110> directions (close packed directions). There are 4 octahedral planes (111), (1$\bar{1}$1), (11$\bar{1}$) and ($\bar{1}$11), 6 <110> directions in each octahedral plane. Each of the directions is common to two octahedral planes, resulting in a total of 12 slip systems. Slip is also possible on (100) along <10$\bar{1}$>. The number of independent slip systems is 12. Slip is illustrated in Figures 7.1 and 7.2.

7.2. Body-Centered Cubic

Slip in BCC occurs on {110} planes (close packed planes), {112} and {123}. Slip direction is <111> (close packed directions).

{110}: There are 12 possible {110} type planes, and for each one there are four slip directions: [1$\bar{1}$1], [$\bar{1}$11], [$\bar{1}$1$\bar{1}$], and [1 $\bar{1}$1]. There are 48 possible combinations of slip plane and slip direction.

{112}: There are 24 possible {112} planes, and for each there are 2 slip directions: [$\bar{1}$11] and [11$\bar{1}$]. There are 48 possible combinations of slip plane and slip direction.

{123}: There are 48 possible planes, and for each there are 2 possible slip directions: [$\bar{1}$11] and [11$\bar{1}$]. There are 96 possible combinations of slip plane and slip direction.

Slip is illustrated in Figures 7.3 and 7.4.

7.3. Hexagonal Close Packed

Pyramidal planes are {hkin}, where n is an integer. Prism type I planes are {h$\bar{h}$00}. Prism type II planes are {hh2$\bar{h}$0}.

Slip [Partridge] in hcp occurs on the basal, pyramidal I and II, and prism I and II, type planes. Slip occurs along prismatic directions <uu $\bar{2u}$ 0> or pyramidal directions <uu $\bar{2u}$ w> on the basal plane {0001}. In vector form, the slip vectors are **c**, **a**, or **c** + **a**. Slip on the basal plane is similar to slip on the octahedral FCC {111} planes. In Figure 7.5 slip on several planes is shown.

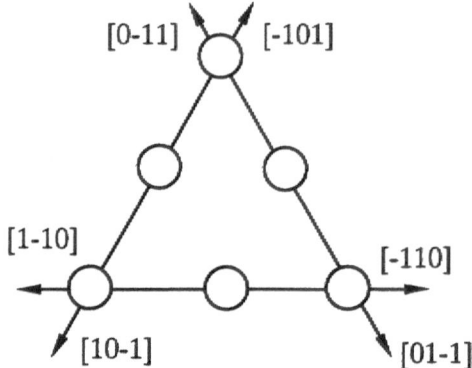

Figure 7.1. FCC slip occurs on close-packed planes in close-packed directions. There are 4 octahedral planes, (111), (1$\bar{1}$1), (11$\bar{1}$), and ($\bar{1}$11), six <110> directions, each one common to two octahedral planes, giving 12 slip systems.

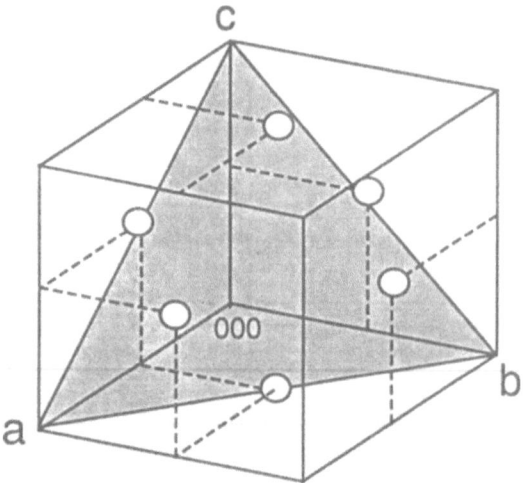

Figure 7.2. The (111) plane in the FCC system is shown shaded.

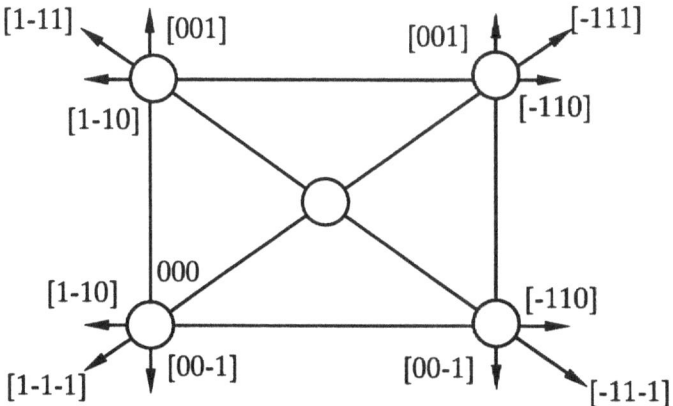

Figure 7.3. BCC slip occurs on close-packed planes in close-packed directions. There are 4 close-packed directions: [111], [$\bar{1}$11], [1$\bar{1}$1], and [11$\bar{1}$] for the (110) plane. Close-packed planes in BCC are {110}, {112}, and {123}.

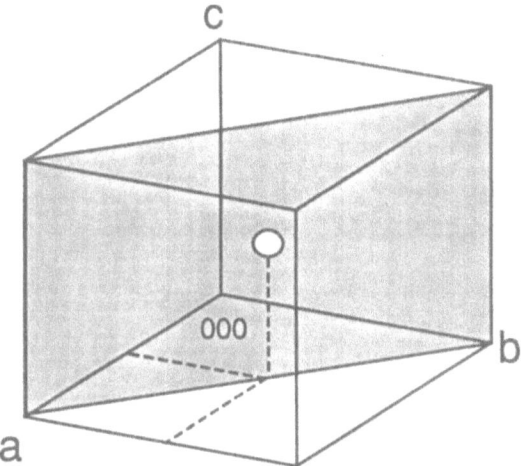

Figure 7.4. The (110) plane in the BCC system is shown shaded.

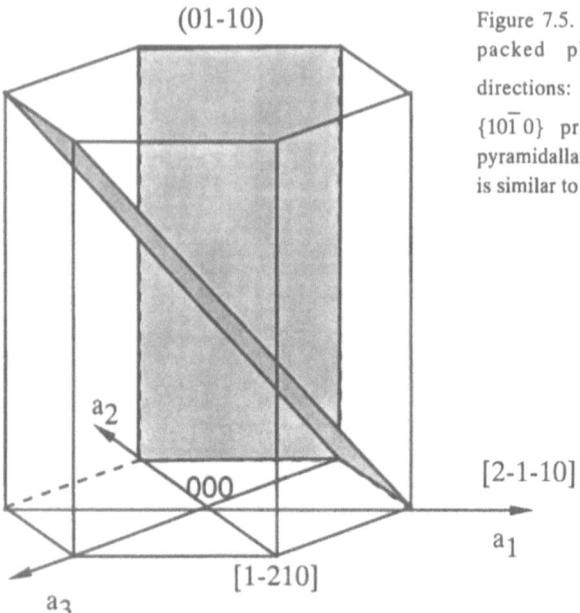

(01-10)

Figure 7.5. hcp slip occurs on close-packed planes in close-packed directions: $<11\bar{2}0>$ (0001), $<11\bar{2}0>$ $\{10\bar{1}0\}$ prism, and $<11\bar{2}3>\{11\bar{2}2\}$ pyramidallanes. Slip on the basal plane is similar to octahedral slip in FCC.

7.4. Miscellaneous Definitions

A notation introduced by Hug et al. [1988] is in use for ordered materials, which has the advantage of producing a concise notation for direction and plane indices:

$<uvw]$ = all permutations of ±u and ±v; w is constant.
$\{hkl)$ = all permutations of ±h and ±k; l remains constant.

The notation is very helpful when superdislocations are present.
    Stacking fault acronyms in use:
        SISF = superlattice intrinsic stacking fault
        SESF = superlattice extrinsic stacking fault
        CSF = complex stacking fault
        APB = antiphase boundary

Tables 7.1–7.3 contain details on the number and type of slip systems in fcc, bcc, and hcp lattices.

Table 7.1. Number of Independent Slip Systems for hcp.

Vector	In (4,4) Notation		Slip Systems		In (3,3) notation
	Direction	Plane	Total	Ind.	
a	$<11\bar{2}0>$	(0001)	3	2	$[110], [0\bar{1}0], [\bar{1}00], [100], [010], [\bar{1}\bar{1}0]$
a	$<11\bar{2}0>$	$\{10\bar{1}0\}$			$[110], [0\bar{1}0], [\bar{1}00], [100], [010], [\bar{1}10]$;
					$(210), (1\bar{1}0), (\bar{1}10), (\bar{2}10), (120), (\bar{1}\bar{2}0)$
	1st order, prism I		3	2	
a	$<11\bar{2}0>$	$\{10\bar{1}1\}$			$[110], [0\bar{1}0], [\bar{1}00], [100], [010], [\bar{1}10]$;
					$(211), (1\bar{1}1), (\bar{1}11), (\bar{2}\bar{1}1), (121), (\bar{1}\bar{2}1)$
	pyramid I		6	4	
a + c	$<11\bar{2}3>$	$\{11\bar{2}3\}$			$[111], [0\bar{1}1], [\bar{1}01], [\bar{1}\bar{1}1], [011], [101]$;
					$(111), (0\bar{1}1), (\bar{1}01), (\bar{1}11), (011), (101)$
	2nd order, pyramid II		6	5	
c	$<0001>$	$\{10\bar{1}0\}$			$(210), (1\bar{1}0), (\bar{1}10), (\bar{2}10), (120), (\bar{1}\bar{2}0)$
	prism I		3	2	
c	$<000l>$	$\{11\bar{2}0\}$			$(110), (0\bar{1}0), (\bar{1}00), (100), (010), (\bar{1}\bar{1}0)$
	prism II		3	2	

After Partridge [1967].

Table 7.2. Number of Independent Slip Systems for FCC, BCC, and hcp.

Crystal System	Systems	Ind.	Total	Crystal System	Systems	Ind.	Total
FCC	$\{111\}<110>$			HCP*	$\{10\bar{1}0\}\{10\bar{1}1\}$		
	$\{100\}<10\bar{1}>$	5	12		$\{11\bar{2}3\}\{11\bar{2}0\}$		
BCC	$\{110\}\{112\},$				$\{0001\}<11\bar{2}0>$		
	$\{123\}<111>$	5	48,		$<0001>$	2–5	24
			48,	*See previous Table for indices in (3, 3)			
			96	notation.			

Table 7.3. Burgers Vectors of Dislocations in hcp Structures.

Vector	Total No. of disl.	Direction Indices of vector (4, 4)	(3, 3)	Magnitude of vector	Energies of dislocations
Perfect:					
a_1, c	6	$\left(\frac{1}{3}\right)<11\bar{2}0>$	[110]	$\lvert \mathbf{a} \rvert$	$\lvert \mathbf{a} \rvert^2$
			[100]		
			[010]		
c	2	$<0001>$	[001]	$\lvert \mathbf{c} \rvert$	$\lvert \mathbf{c} \rvert^2 = \left(\frac{8}{3}\right)\lvert \mathbf{a} \rvert^2$
c + a	12	$\left(\frac{1}{3}\right)<11\bar{2}3>$	[111]		
			[011]	$\sqrt{\lvert \mathbf{c} \rvert^2 + \lvert \mathbf{a} \rvert^2}$	$\left(\frac{1}{3}\right)\lvert \mathbf{a} \rvert^2$
Imperfect:					
$\left(\frac{1}{3}\right)(2\mathbf{a}_1 + \mathbf{a}_2)$ $= \mathbf{a'}$		$\left(\frac{1}{3}\right)<10\bar{1}0>$	$\left(\frac{1}{3}\right)[\bar{1}10]$		
			$\left(\frac{1}{3}\right)[210]$	$\dfrac{\lvert \mathbf{a} \rvert}{\sqrt{3}}$	$\left(\frac{1}{3}\right)\lvert \mathbf{a} \rvert^2$
$\dfrac{c}{2}$	4	$\left(\frac{1}{2}\right)<0001>$	$\left(\frac{1}{2}\right)[001]$	$\dfrac{\lvert \mathbf{c} \rvert}{2}$	$\left(\frac{2}{3}\right)\lvert \mathbf{a} \rvert^2$
$\dfrac{c}{2} + \mathbf{a'}$		$\left(\frac{1}{4}\right)<20\bar{2}3>$	$\left(\frac{1}{4}\right)[423]$		
			$\left(\frac{1}{4}\right)[\bar{2}23]$	$\sqrt{\dfrac{\lvert c \rvert^2}{4} + \dfrac{\lvert a \rvert^2}{3}}$	$\lvert \mathbf{a} \rvert^2$

After Partridge [1967].

8
Projections

8.1. Introduction

The calculation of projections of a plane onto a reference plane parallel to it can be done in a straightforward way using vectors. The utility of these calculations in this form is, first, that hand calculation is simple, and, second, that the vector equations can be easily translated into a computer program if desired.

Applications of the procedure include convergent beam analyses, particularly for symmetry determination, calculation of Brillouin zones in reciprocal space, and construction of three-dimensional models of a structure or of reciprocal space.

The stereographic projection has wide use in x-ray diffraction analyses, but its use in electron diffraction is not as prevalent, the applications including orientation relationship determination, twin analysis, determination of accessibility of planes and zones experimentally using a given specimen holder.

More general projections are in use as well, and Verma and Srivastava [1981] provide an extensive description of such projections.

8.2. Direct Lattice Projections

When a crystal structure is viewed along some arbitrary direction, the atoms in each plane perpendicular to that direction may or may not lie exactly above each other. In FCC, for example, the stacking along [111] is ABC, i.e., three equivalent planes the atoms of which occupy specific positions shifted with respect to the first plane designated as A. Figure 8.1 illustrates the stacking of the adjacent layers. If the x and o atom positions in each plane were shifted by some amount so that all were to lie over the center marked by the square, then the stacking would be ...AAA... . This shift can be calculated by noting the geometry of the structure.

Figure 8.2 is a projection of an arbitrary plane in a structure in the direct lattice.

D = vector defining the direction **D** perpendicular to the plane P.

d = vector to the atom position with coordinates [u,v,w].

t = vector from the origin to a point on P where the projection of **d** terminates.

R = vector projection of **d** onto **D**.

θ = the angle between **d** and **R**.

Figure 8.1. Stacking of FCC [111].
x, o represent different levels.

89

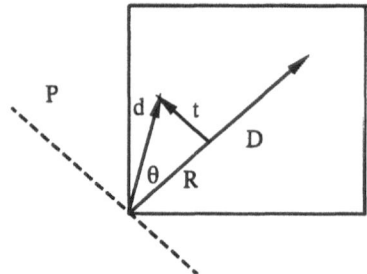

Figure 8.2. Geometry of arbitrary plane and the projection onto a reference plane.

From the geometry shown in Figure 8.2,

$$t = d - \frac{(d \cdot D)}{|D|^2} D. \tag{8.1}$$

It is convenient to write t as a fraction of a vector to an atom position in the reference plane P,

$$t = f \, r_0. \tag{8.2}$$

In terms of indices,

$$t = f \, [\, u,v,w \,] \, r_0. \tag{8.3}$$

Hence, t is the projection of an atom position onto P and is written in terms of an atom position in plane P.

Example: Cubic system.

1. For $D = [110]$, $d = [100]$:

$$t = [100] - \frac{1}{2} [110] = \frac{1}{2}[1\bar{1}0]. \tag{8.4}$$

Hence, $f = 1/2$ and $r_0 = [1\bar{1}0]$.

2. For $D = [111]$, $d = [1\bar{1}1]$:

$$t = [1\bar{1}1] - \frac{1}{3}[111] = \frac{2}{3}[1\bar{2}1]. \tag{8.5}$$

Hence, f is 2/3, and r_0 is $[1\bar{2}1]$.

The utility of the equations lies in the capability to calculate the projection of atom positions or planes onto a reference plane. So, one can construct a projection of a structure onto a reference plane showing atom positions for all the repeat layers for a simple or a complex motif.

8.3. Reciprocal Space Projections

By following a similar procedure, one can construct projections of reciprocal lattice points, from which diffraction patterns can be constructed corresponding to the usual selected area diffraction patterns (SADP) described in Chapter 3, as well as the reciprocal lattice layers, which lie above the plane in reciprocal space containing the

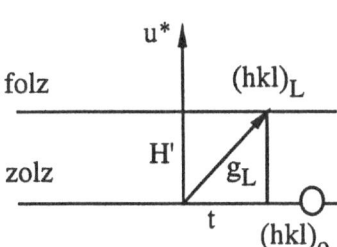

SADP, the so-called "holz" (higher order Laue zones) planes described more fully in Chapter 10.

Proceeding as in the previous section, from the geometry shown in Figure 8.3,

$$t = g_L - u^* \, H', \tag{8.6}$$

where g_L = the lattice vector from the origin in the zolz to some $(hkl)_L$ lattice point in zone L; H' = reciprocal lattice separation of Laue zones in units of reciprocal length; u^* = the vector perpendicular to zolz and parallel to u. It represents a plane.

The unit vector

$$\hat{u}^* = \frac{u^*}{|u^*|} \tag{8.7}$$

is used in the following equations.

For the cubic case

$$\hat{u}^* \, H' = u^* \, \frac{H}{|u^*|} = \frac{1}{u^2 + v^2 + w^2} \, . \tag{8.8}$$

The equation for t can be rewritten as

$$t = g_L - u^* \, \frac{H}{|u^*|} \, , \tag{8.9}$$

which allows calculation of t using known indices. Now t lies in the reference plane, i.e., the zero order Laue zone or zolz, along the vector connecting the origin and a zolz plane $(hkl)_0$. Hence, $|\,t\,|$ can be represented as some fraction of the g_0 and the direction is that of g_0. This is done by writing t as

$$t = f \, g_0. \tag{8.10}$$

In terms of indices

$$t = f \left[(hkl)_L - (u^*v^*w^*)\frac{H}{|u^*|} \right] .$$ (8.11)

Example for cubic system:
    Given the zone $u = [110]$, for the first order Laue zone or folz $g \cdot u = 1$. A plane in the folz is (100) since this satisfies the zone law. The t vector is

$$t = (100) - \frac{1}{2}(110) = \frac{1}{2}(1\bar{1}0).$$ (8.12)

So this means that the projection of the (100) lattice point onto the zolz is located along the $(1\bar{1}0)$ g vector and is at a distance of $g/2$. With a magnitude and a direction established, the locations of all the planes in the folz are established since the folz pattern is the same as the zolz except for a general shift of the origin resulting from the structure. The objective of locating the folz pattern relative to the zolz has been accomplished. Identification of the indices associated with each beam is described in Chapter 10. A table of equations for the H for each of the seven crystal systems is also included in Chapter 10 (Table 10.12).

8.4. The Stereographic Projection

Since angles between planes in a diffraction pattern correspond to angles in the direct lattice, they provide a convenient link between the direct and reciprocal spaces. A means for representing the angles between planes that has angles as variables, as opposed to repeat distances, would thereby allow comparisons between the lattice and the diffraction pattern. The stereographic projection is such a geometric means, allowing direct experimental measurements from the diffraction pattern to be applied to obtain information about the direct lattice.
    Place the center of a unit cell at the center of a sphere of radius r_s, as in Figure 8.4. To the planes of the unit cell attach normal vectors of a length such that these vectors contact the surface of the sphere. The result is a sphere with points on the surface separated by angles corresponding to the angles between normals and, therefore, between planes. The task is to project these points systematically onto a two-dimensional plane.

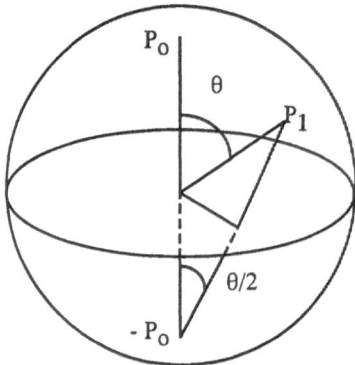

Figure 8.4. Geometry of points in the stereographic projection. The reference plane is defined by the plane that is perpendicular to point P_0 and that intersects the sphere producing the great circle at the equator of the sphere.

Choose any point P_0 on the surface and let it represent a normal to a plane at the center of the sphere. Allow the plane perpendicular to P_0 to intersect the surface of the sphere, producing an "equator" or great circle (radius the same as the sphere). Choose any other point P_1 on the surface and located at some angle, θ, from P_0. P_1 is also assumed to be a normal to a plane located at the center of the sphere. Geometrically, it can be shown (see Figure 8.4) that the angle that P_1 makes with $-P_0$ is $\theta/2$. If one draws a line from the center of the sphere along the equatorial plane to the point where the line connecting $-P_0$ and P_1 intersects the equatorial plane, one finds that the distance is given by

$$r_{01} = r_s \tan \frac{\theta}{2}, \tag{8.13}$$

where r_s is the radius of the sphere. See Figure 8.4. This equation is the basis for locating projected points on the equatorial plane corresponding to the point (or pole) P_0. Thus for any point lying on the surface of the sphere, its position can be projected onto the equatorial plane and its location can be found in terms of the angle it makes with the vector normal to the equatorial plane.

An alternative way of looking at this projection is to take the equatorial plane and place a copy of it outside the sphere and in contact only at the normal vector point P_0. By extending the vector connecting $-P_0$ and P_1 beyond the sphere until it now intersects the plane, an enlarged map of points on the surface of the sphere can be generated. The positions of these points are determined by the angle from the normal vector, θ, and the azimuthal angle of rotation about the normal vector, ϕ. Hence, one now has the means for representing the angles between the normals to planes in a crystal without direct reference to lattice repeat distances, although these dependences are built into the angles through the equations given in Chapter 5.

To make the connection with crystal systems explicit, define P_0 to be either normal to the crystal plane (hkl) or normal to the direction [uvw]. In the first case, one has a plane projection; in the second case, one has a pole projection of directions. Because, in general, poles and normals to planes may not be parallel, the two projections may not coincide, except for the cubic system. Stereograms for the cubic system are universal for any cubic lattice parameter.

Equation (8.13) is used to generate a Wulff net (see Figure 8.5), which is a grid of equal angles, in increments of $2°$. The grid, however, is not linear in angle, i.e., the distance between angles on the grid increases as one moves out from the center.

The stereographic projection conventions are that the planes lying in the same hemisphere as P_0 are included in the projection. Planes below the P_0 hemisphere are not included. By convention the two axes at $90°$ to each other and which lie in the P_0 plane are labeled N-S and W-E, as indicated in Figure 8.6.

Given the pole P_0, its plane, and a pole P_1, connect P_0 and P_1 on the sphere's surface and continue the line all the way around until it reconnects to P_0. The result is another "equatorial" curve or great circle. Its projection onto the reference plane of P_0 is a straight line. If point P_1 is moved onto a position lying on the reference great circle the axis produced by the equatorial and its projected line is defined as the N-S line.

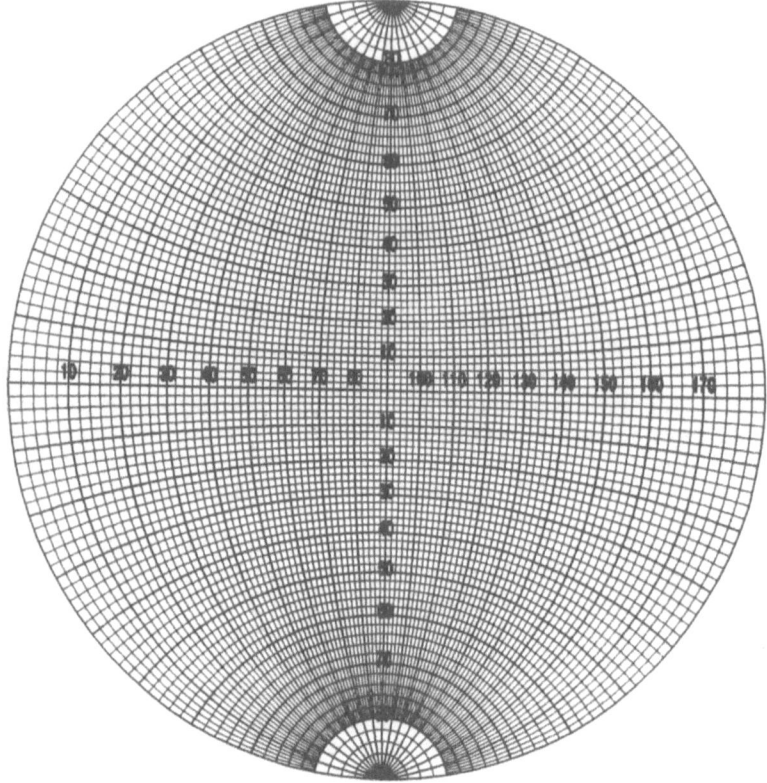

Figure 8.5. The Wulff net used for stereographic projection analyses.

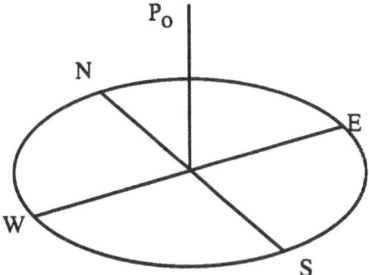

Figure 8.6. Defining geometry of the N-S and the W-E axes lying in the reference projection plane of the pole P_0.

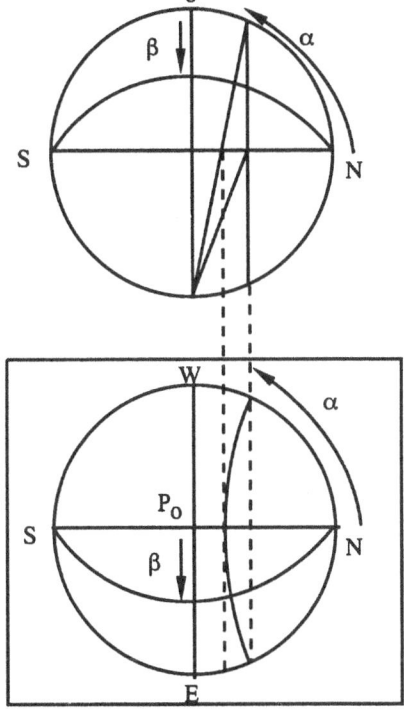

Figure 8.7. Geometry of projections of latitudes at angle α and longitude at angle β.

Choose a second point P_2 on the equatorial plane of P_0, which is now called the reference great circle, and place this point 90° from the N-S line. The projection of the great circle connecting this point and P_0 is also a straight line. Define this line to be the W-E axis.

In order to conceptually make the projection easier to follow, points on the surface of the sphere are located by reference to angles with respect to the N or the S pole and angles with respect to the W or the E pole. The convention for the azimuthal angle places the 0° position at N or at S, and the 90° position at W or at E, all in the reference plane of P_0.

Consider great circles on the sphere arranged so that the circles have the N-S axis in common. Projections of these great circles onto the reference plane produce curved lines that meet at N and at S. These lines correspond to the *lines of longitude* used to locate movement east or west on a globe.

Circles on the sphere perpendicular to the N-S axis and having the N-S axis as center are not great circles except at the "equator" 90° from the N or S. The projection of these lines onto the reference plane produces hyperbolas with N or S as focus. These lines correspond to *latitudes* used to measure distances from the north or the south poles. Hence, in the projection, there are longitudes and latitudes. Rotation about the projection pole corresponds to a change in latitude, whereas rotation about the E-W axis corresponds to change in longitude. The geometry is illustrated in Figure 8.7.

Detailed descriptions and explanations of the stereographic projection and other projections are presented in Cullity [1978], Verma and Srivastava [1981] and Wahlstrom [1983].

8.5. Grid Projections

Grid projections for cubic (110) plane with rectangles $1 \times \sqrt{2}$ on a side, for orthohexagonal with rectangles $1 \times \sqrt{3}$ on a side, and for hexagonal (0001) or cubic (111) planes are given in Figures 8.8, 8.9, and 8.10. These projections are especially useful for sketching projections of reciprocal space, for indexing holz patterns (see Chapter 10), and for sketching direct lattice projections.

Figure 8.8. Grid pattern with rectangle side ratio of $1:\sqrt{2}$.

Figure 8.9. Grid pattern with rectangle side ratio of $1:\sqrt{3}$.

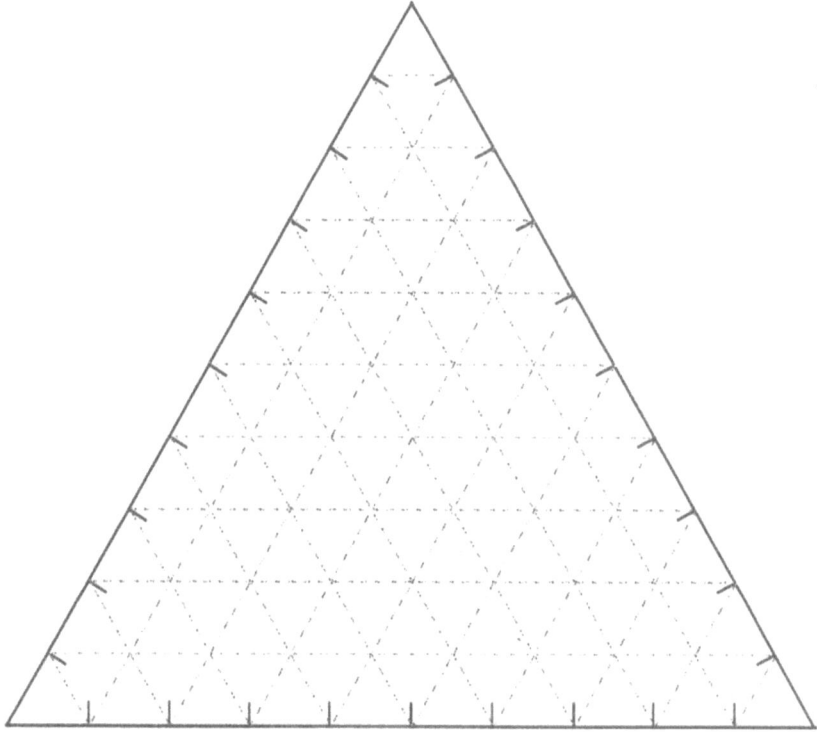

Figure 8.10. Triangular grid of equilateral triangles. This can be used as a template for hexagonal coordinates.

9
Structure Symbols

9.1. Crystal Designations

9.1.1. Introduction

The hierarchy of the crystal families is illustrated in Figure 9.1. In this diagram, there are seven crystal systems that include subdivisions of Bravais lattices, point groups, glide planes, screw axes, and space groups. The rationale for this hierarchy is discussed in detail in the International Crystallography Tables [1983]. The symbols described in the following sections are based on this hierarchy. Other symbol systems are in use, as well, but this hierarchy of systems to lattices to groups is the most widely used in materials science.

Among the seven crystal systems there are 14 possible ways to define a space lattice. These possibilities constitute the 14 Bravais lattices, as shown in Figure 9.2. All crystal structures can be reduced to one or more of these 14 Bravais lattices, except for the pentagonal structures. These require special handling, as described in Chapter 16. The characteristics of each of the crystal systems are listed in Table 9.1.

9.1.2. Equivalent points

If an object is present at xyz in the unit cell, then the symmetry of the unit cell will produce the object at coordinates $x_2y_2z_2,...$. Points related by symmetry operations are equivalent points [Verma and Srivastava, 1981].

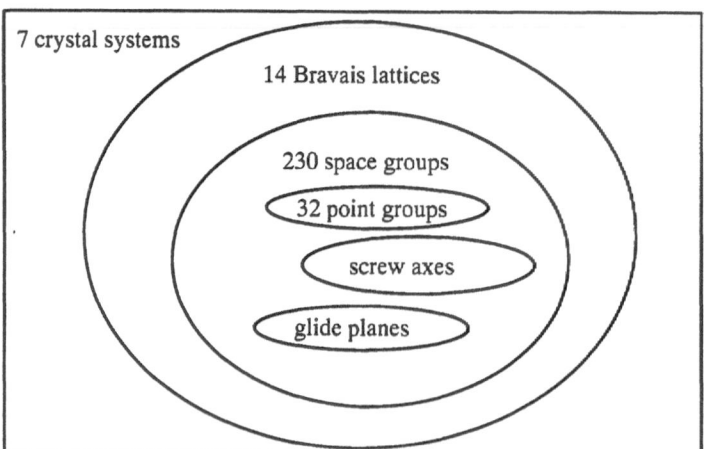

Figure 9.1. The hierarchy of the crystal systems.

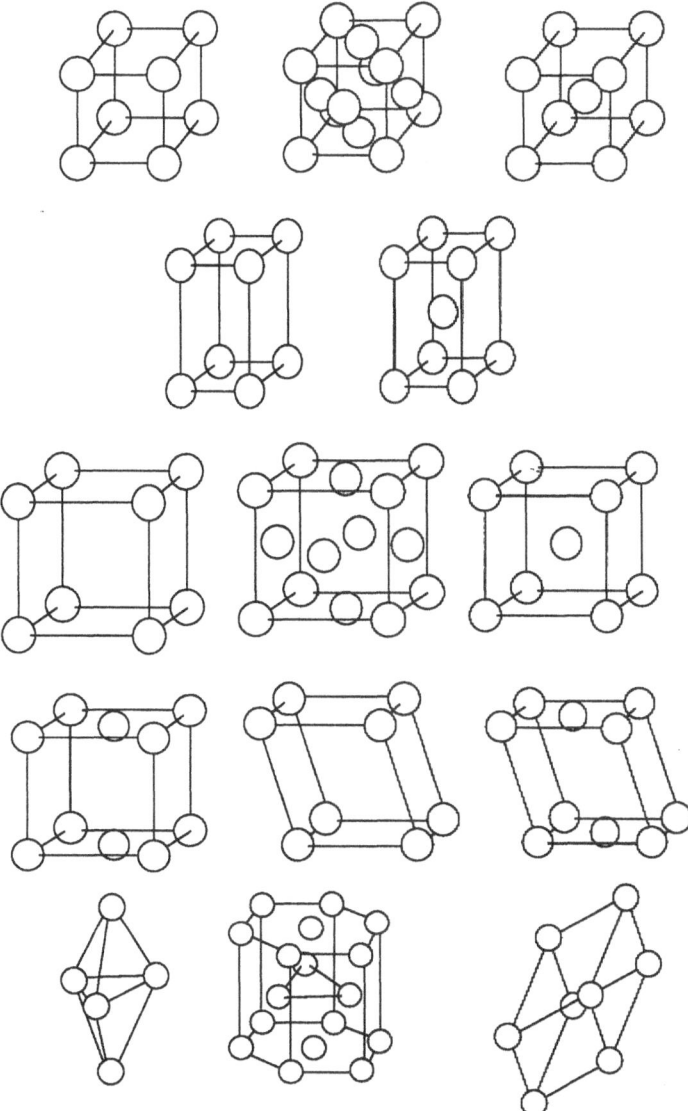

Figure 9.2. The 14 Bravais lattices. First row: simple, face- and body-centered cubic; second row: simple and body-centered tetragonal; third row: simple, face-, body-centered orthorhombic; fourth row: end-centered orthorhombic, simple and end-centered monoclinic; fifth row: trigonal, hexagonal, triclinic.

When a center of symmetry is present with an inversion center at 000, the point xyz has the equivalent point $-x, -y, -z$. For a center at x_c, y_c, z_c, if the origin is shifted to this point, then the coordinates of xyz become $x - x_c, y - y_c, z - z_c$. The inversion center becomes $x_c - x, y_c - y, z_c - z$. Translating the inverted point coordinates to the original origin, i.e., transforming back to 000, gives $2x_c - x$, $2y_c - y, 2z_c - z$. So for an inversion center at x_c, y_c, z_c we have

point: equipoint:
xyz $2x_c - x,\ \ 2y_c - y,\ \ 2z_c - z.$

As a second example, consider the case of a 2–fold axis. For the axis oriented along the z axis such that the axis is located in the xy plane at the point $x_0, y_0, 0$, then

point: equipoint:
xyz $2x_0 - x,\ \ 2y_0 - y,\ \ z.$

For a mirror plane perpendicular to z and the mirror at z_m, we have

point: equipoint:
xyz $x,\ \ y,\ \ 2z_m - z.$

For a 3-fold axis oriented along the z axis and with the axis at 0, 0, 0, we have

point: equipoint:
xyz $\bar{y},\ \ x + \bar{y},\ \ z;\ \ \ \ \bar{x} + y, \bar{y}, z.$

For other axes, the point-equipoint relationships, using the bar notation for negative coordinates, are

$\bar{3}$: $y, y + \bar{x}, \bar{z};\ \ \ x + \bar{y}, y, \bar{z}$
4: $\bar{y}, x, z;\ \ \ \bar{x}, \bar{y}, z;\ \ \ y, \bar{x}, z$
$\bar{4}$: $y, \bar{x}, \bar{z};\ \ \ x, y, \bar{z};\ \ \ \bar{y}, x, \bar{z}$
6: $x + \bar{y}, x, z;\ \ \ \bar{y}, x + \bar{y}, z;\ \ \ \bar{x}, \bar{y}, z;\ \ \ \bar{x} + y, \bar{y}, z;\ \ \ y, \bar{x} + y, z$
$\bar{6}$: $\bar{x} + y, \bar{x}, z;\ \ \ y, \bar{x} + y, \bar{z};\ \ \ x, y, \bar{z};\ \ \ x + \bar{y}, y, \bar{z};\ \ \ \bar{y}, x + \bar{y}, \bar{z}$

For screw axes (see Figure 9.3), the relationships are

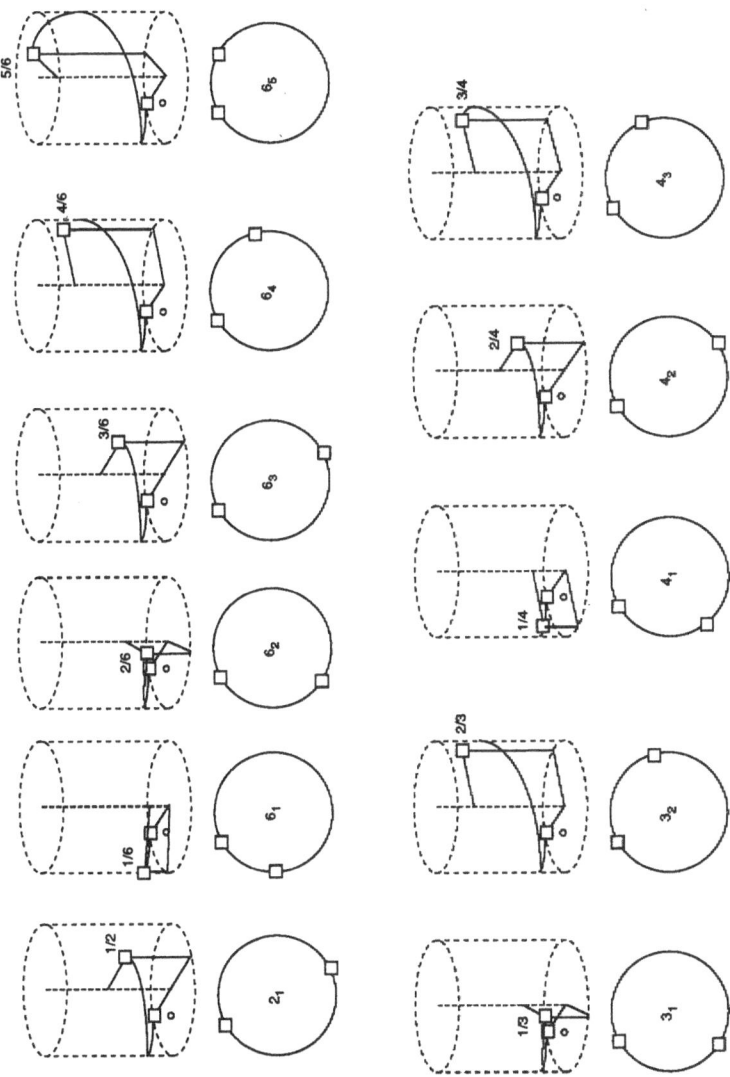

Figure 9.3. The geometries defining the various screw axes are indicated.

2_1: xyz $\quad\quad \bar{x}, \bar{y}, \frac{1}{2} + z$ $\quad\quad\quad\quad$ 6_1: xyz $\quad x + \bar{y}, x, \frac{1}{6} + z$

$\quad\quad\quad\quad\quad\quad\quad\quad\quad\quad\quad\quad\quad\quad\quad$ 6_2: xyz $\quad \bar{y}, x + \bar{y}, \frac{2}{6} + z$

3_1: xyz $\quad\quad \bar{y}, x + \bar{y}, \frac{1}{3} + z$

$\quad\quad\quad\quad\quad\quad\quad\quad\quad\quad\quad\quad\quad\quad\quad$ 6_3: xyz $\quad \bar{x}, \bar{y}, \frac{3}{6} + z$

3_2: xyz $\quad\quad \bar{x} + y, \bar{y}, \frac{2}{3} + z$

$\quad\quad\quad\quad\quad\quad\quad\quad\quad\quad\quad\quad\quad\quad\quad$ 6_4: xyz $\quad \bar{x} + y, \bar{y}, \frac{4}{6} + z$

4_1: xyz $\quad\quad \bar{y}, x, \frac{1}{4} + z$ $\quad\quad\quad\quad$ 6_5: xyz $\quad y, \bar{x} + y, \frac{5}{6} + z$

4_2: xyz $\quad\quad \bar{x}, \bar{y}, \frac{2}{4} + z$

4_3: xyz $\quad\quad y, \bar{x}, \frac{3}{4} + z$

Glide elements are listed in Table 9.2. Glide plane relationships are:

a $\quad$ xyz $\quad\quad \frac{1}{2} + x, y, \bar{z}$

b $\quad$ xyz $\quad\quad x, \frac{1}{2} + y, \bar{z}$

c $\quad$ xyz $\quad\quad \frac{1}{2} + x, \frac{1}{2} + y, \bar{x} + z$.

9.2. Strukturbericht Symbols

The symbol consists of a letter and a number [Barrett and Massalski 1980; Pearson, 1972]. The letters serve to categorize types of structures; the numbers are used for designating various types within a letter category. See Table 9.3 for definitions of the symbols.

9.3. Pearson Symbols

The Pearson symbol consists of a lower case letter symbolizing the Bravais lattice system, a letter symbolizing the centering, and a number representing the number of atoms in the unit cell [Pearson 1972; Villars and Calvert, 1985]. See Table 9.4 for definitions of the symbols.

9.4. Symmetry Symbols

9.4.1. Operational definitions

The symbols [ITC, 1983; Ladd and Palmer, 1988] are based on the 32 point symmetry groups i.e., the symbols arise from the various ways in which equivalent points can be arranged with respect to a reference point. There are a number of definitions that are required in explaining the point groups. The most commonly needed definitions are as follows:

basis = the group of atoms associated with a lattice point. The vector associated with the basis is $x_j a_1 + y_j a_2 + z_j a_3$, where x_j, y_j, z_j are the fractional coordinates of the atom j in the basis referred to the origin of the unit cell.

primitive cell = the axes are chosen such that there is only one lattice point per cell.

unit cell = cell by which the entire lattice can be reproduced.

symmetry element = an operation by which equivalent points are brought into coincidence.

equivalent point = position in the crystal unit cell related to another point in the unit cell by a symmetry element.

crystallographic axes = a, b, c; a set of noncoplanar vectors.

9.4.2. Macroscopic symmetry elements

Macroscopic symmetry elements are elements related to points and are two dimensional in nature. The macroscopic symmetry elements are mirror, m, and the rotational axes are as follows:

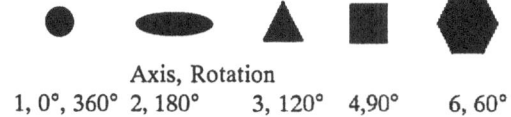

Axis, Rotation
1, 0°, 360° 2, 180°   3, 120° 4,90°   6, 60°

These rotations define symmetry axes.

Symmetry plane = reflections of equivalent points through a plane.

Inversion center = every point on one side is carried through a center point to the opposite side producing an equivalent point. This is equivalent to a π rotation followed by a reflection in the plane perpendicular to the rotation axis.

Rotation-inversion = combined rotation then inversion or inversion then rotation. This operation translates equivalent points to points attainable by reflection in a mirror horizontally placed with respect to the vertical rotation axis.

2-fold rotation-inversion is equivalent to a reflection, and a 1-fold inversion is a center of inversion.

By combining the five basic symmetry elements, the 32 point groups are generated. These are listed with their symbols in Tables 9.5 and 9.6.

9.4.3. Space group symbols

Macroscopic symmetry allows identification of 32 crystal point groups and seven crystal systems. The symmetry elements are defined in terms of points and manipulations of these points. Hence, the macroscopic symmetry elements are equivalent to point groups. When the symmetry of the arrangement of atoms in a lattice is considered, the array of symmetry elements on a space lattice is called a space group. There are 230 possible space groups in classical crystallography. The addition of 5-fold symmetry in nontranslationally periodic structures increases this number. For further discussion of the 5-fold and higher elements. See Chapter 16.

The symmetry elements operate on equivalent points within the unit cell of the crystal lattice, and so are in general three dimensional.

When a translation vector and a reflection vector are added, the result is a glide plane. The restriction on the vectors are that the reflection vector must be perpendicular to the translation vector, and the translation vector must lie in the mirror plane. Thus in lattices with mirror planes, glide planes may appear. Glide planes are divided into three categories, depending on the nature of the translation vectors [ITC, 1983].

If a translation is combined with a rotation the result is a screw symmetry element. There are screw axes for 2, 3, 4, 6 -fold axes, as shown in Figure 9.3.

9.4.4. International symbols

There have been a number of symbols used for space groups, and of these only two or three remain in use. The international symbol is the same as the Hermann-Mauguin symbol. Others in use are the Schoenflies, Jagodzinski-Wyckoff, Ramsdell and Zhdanov notations. The last three are discussed in Sections 9.4.6-9.4.8.

9.4.5. Schoenflies symbol

The Schoenflies symbol [ITC,1983; Vainstein, 1979] arises from the symbols used to designate various rotation groups: D_h = dihedral, C_n = cyclic, T = tetrahedral, O = octahedral, and K = icosahedral groups. The last three are related to permutation groups.

Symbol: $D_a{}^b$, where D = rotation group; a = point group; b = space group number (optional).

See Table 9.6 for the Schoenflies symbols for the 32 point groups.

9.4.6. International or Hermann-Mauguin symbol

This symbol [ITC, 1983] has two forms generally, some groups having more than two as a result of special situations. The short form consists of the type of lattice and the symmetry elements of the principal directions for the crystal system. Thus, the symbol is

$$Xx_1x_2x_3,$$

where X = Capital letter: P = simple lattice or primitive lattice; A, B, C = lattice centered on the A, B, or C face; F = face centering; I = body centering. For hexagonal, the letter symbols are H = axial ratio a/b = $\sqrt{3}$; R = trigonal with rhombohedral axes, i.e., with equal axes and equal angles between axes. Note that in some references the symbol C is used to designate an ortho-hexagonal lattice with centering on the C face and with a/b = $1/\sqrt{3}$. This symbol is not currently in use because of possible confusion with C-centering symbols in the orthogonal systems.

The short form of the International symbol also may consist of the system symbol and a single letter or all three letters, depending on the system.

The long form of the International symbols includes three axes (principal, secondary, tertiary) for every crystal system: $x_1 x_2 x_3$ = symmetry elements of special directions. For the seven systems the axes are:

triclinic: x_1 = one of the lattice axes; (long form: x_1 = a; x_2 = b; x_3 = c.)

monoclinic: the axis normal to the other two; (long form: x_1 = b; x_2 = a; x_3 = c.)

orthorhombic: x_1 = a; x_2 = b; x_3 = c.

tetragonal: x_1 = principal axis parallel to the 4-fold axis = c; x_2 = axis 90° to x_1 = a; x_3 = axis 45° to x_2, i.e., the axis that lies along the <110> direction.

cubic: x_1 = [100] = a; x_2 = [111]; x_3 = [110].

hexagonal: x_1 = axis parallel to the 3- or 6-fold axis = c; x_2 = axis 90° to x_1 = a; x_3 = [110].

trigonal: x_1 = axis parallel to 3-fold axis = c; (long form: x_1 = c; x_2 = a; x_3 = [110].)

The characteristics of the seven systems are listed in Table 9.1.

The full details of each of the 230 space groups appear in the International Tables for Crystallography, Vol A [ITC, 1983], or the International X-Ray Tables [IXTC, 1959].

The use of rhombohedral, trigonal, and hexagonal names, and the hexagonal and rhombohedral settings for lattice axes can lead to some confusion. The hexagonal crystal family is divided into hexagonal (6-fold symmetry) and trigonal systems (3-fold symmetry). The trigonal system is further divided into two settings, one with hexagonal coordinates, the other with rhombohedral coordinates. The unit cells in the rhombohedral and the hexagonal coordinate settings are related: the rhombohedral transforms into the hexagonal triple obverse cell, which is the preferred cell and axes [ITC, 1983].

Table 9.6 lists the crystal system, symmetry elements, both full and abbreviated, and other symbols used for the element. [Barrett and Massalski, 1980, Brown and Forsyth, 1979, Vainstein, 1979]. The symmetry formula column gives the axes, centers and planes together with the number of the elements: L = axes; C = center; P = plane of symmetry; the number of elements is the number before the symbol. The superscript refers to the -fold of the axis, the subscript to the inversion axis.

The common strukturbericht symbols, with Pearson symbol, archtype, and space group are listed in Table 9.7. Tables 9.8 through 9.10 contain the same sets of symbols arranged by archtype, by Pearson symbol, and by space group.

9.4.7. Jagodzinski-Wyckoff notation

The impetus for this notation [Pearson, 1972; Parthe, 1966] is the existence of variations in stacking which produce complex structures. For detailed explanation see Parthe [1964] or the International Tables [ITC, 1983; ITXC, 1959].

h = unit slab whose neighboring slabs, above and below, are displaced in the same direction; hexagonal reference, e.g., A3, B4, and C14 types.

c = unit slabs whose neighboring slabs show sideways displacements different from h; cubic reference, e.g., A1, B3, C15 types.

Examples of stacking sequences:

B4: h; B3: c; B5: hc; B6: hcc; B7: hchcc.

The number of actual layers may be equal to or twice the number of symbols employed.

9.4.8.   Ramsdell notation

This notation [Pearson, 1972; Parthe, 1966] seeks to do the same thing as the Jagodzinski-Wyckoff notation, but it uses number-letter combinations instead of h and c.:

(number)(letter) = {number of layers in the sequence}{structure type}.

Example: Sm: ABABCBCAC = hhc = 9R (h = ABA, h = BAB, c = ABC).

The sequence is hhchhc, and there are nine layers in a rhombohedral structure.

9.4.9.   Zhdanov notation

This notation [Pearson, 1972; Parthe, 1966] is a numerical variation of the Jagodzinski-Wyckoff notation. It uses numbers to represent the number of types of stacking present:

Example: Sm: hhc = (12)
          SiC(I): hcchc = (32).

A new section starts with each h.

Table 9.1a. Characteristics of The Seven Crystal Systems.

System	Point Groups	Unit Cell	Standard Axes
Triclinic	1, $\bar{1}$	$a \neq b \neq c, \alpha \neq \beta \neq \gamma$	**a**,**b**,**c** not coplanar
Monoclinic	2, m, 2/m	$a \neq b \neq c; \alpha = \gamma = 90°$ $90° < \beta < 120°$	**b** chosen parallel to 2-fold axis, or perpendicular to m; **a**, **c** perpendicular to **b** and smallest lattice vectors
Orthorhombic	mm, 222, mmm	$a \neq b \neq c,$ $\alpha = \beta = \gamma = 90°$	**a**, **b**, **c** oriented along the 2-fold axes, or **a**, **b**, **c** oriented parallel to the perpendiculars to the mirror planes
Tetragonal	4, $\bar{4}$ 4/m, $\bar{4}$2m, 4mm, 422, 4/mmm	$a = b \neq c,$ $\alpha = \beta = \gamma = 90°$	**c** oriented parallel to the 4 or $\bar{4}$ -fold axis; **a**, **b** oriented perpendicular to **c** and the smallest lattice vectors.
Hexagonal	6,$\bar{6}$, 6/m, $\bar{6}$2m, 6mm, 622, 6/mmm	$a = b \neq c,$ $\alpha = \beta = 90°; \gamma = 120°$	**c** parallel to the 6 or $\bar{6}$-fold axis; **a**,**b** perpendicular to c and smallest lattice vectors.
Trigonal:	3, $\bar{3}$, 3m, 322, 3m		
Hexagonal setting:		$a = b \neq c$ $\alpha = \beta = 90°; \gamma = 120°$	**c** parallel to the 3- or 3-fold axis; **a**, **b** smallest lattice vectors perpendicular to **c** and related to each other by 3- or -3-fold symmetry
Rhombohedral setting:		$a = b = c;$ $\alpha = \beta = \gamma \neq 90°$ and $< 120°$	**a**, **b**, **c**, the smallest noncoplanar lattice vectors related to each other by 3- or 3-fold symmetry; **a**, **b**, **c**, not perpendicular to each other and are not parallel to the 3- or -3-fold axis.
Cubic	23, m3, $\bar{4}$3m, 43, m3m	$a = b = c;$ $\alpha = \beta = \gamma = 90°$	**a**, **b**, **c** parallel to the three 2-fold axes (23 and m3), or $\bar{4}$-fold axis (43m), or 4-fold axis (43, m3m).

Adapted from ITC [1983], Barrett and Massalski, [1980], and Kitaigorodskii [1955].

Table 9.1b. Characteristics of the Crystal Systems.

Lattice Type	Crystal System	Minimum Symmetry	Point Groups	Orientation of Axes	Crystal Axes and Angles	Space Group Symbols	Unit Cell Transformations Possible
P	Triclinic	1 or $\bar{1}$	$1, \bar{1}$	unspecified	$a \ne b \ne c$, $\alpha \ne \beta \ne \gamma$	$P1, P\bar{1}$	----
P,C	Monoclinic	mirror, one 2 or $\bar{2}$	$2, m, \frac{2}{m}$	c parallel to 2	$a \ne b \ne c$, $\alpha = \gamma = 90°$, $\beta \ne 90°$	$P[010]$, $C[010]$ b axis unique	P,C or P,A or P,I Depending on unit cell
P,C, I,F	Orthorhombic	three 2 or $\bar{2}$	222, mm2, mmm	a,b,c parallel to 2	$a \ne b \ne c$, $\alpha = \beta = \gamma = 90°$	Pabc, Iabc, Cabc, Fabc	I to P, or F to P
P,I	Tetragonal	one 4 or $\bar{4}$	$4, \bar{4}, \frac{4}{m}$, 422, 4mm, $\bar{4}2m$, $\frac{422}{mmm} = 4/mmm$	c parallel to 4	$a = b \ne c$, $\alpha = \beta = \gamma = 90°$	$Pc,a[110]$, $Ic,a[110]$	P to C or I to F
P,I,F	Cubic	four 3 or $\bar{3}$	23,432, $\frac{2}{m}\bar{3} = m3$, $\bar{4}3m, m3m$, $\frac{4}{m}\bar{3}\frac{2}{m}$	a,b,c parallel to 4	$a = b = c$, $\alpha = \beta =$, $\gamma = 90°$	$Pa,[111], [110]$, $Fa,[111], [110]$, $Ia,[111], [110]$	I to P or F to P
P, R	Hexagonal	one 6 or $\bar{6}$	$6, \bar{6}, 622$, 6mm, $\bar{6}m2$, $\frac{6}{m}, \frac{622}{mmm} = 6/mmm$	c parallel to 6 [111]	$a = b \ne c$, $\alpha = \beta = 90°$, $\gamma = 120°$	$Pc,a[110]$	hcp to orth C; hcp to triple hex H
R	Trigonal	one 3 or $\bar{3}$	$3, \bar{3}, 32$, 3m, $\bar{3}\frac{2}{m} = \bar{3}m$	c parallel to a=b=c, [111]	$a = b = c$, $\alpha = \beta = \gamma \approx 90°$	Pc,a, Rc,a	hex to triple rhom D; R to H

P = primitive, I = body-centered, F = face-centered, C = C base-centered, R = rhombohedral, trigonal

Table 9.2. Glide Elements.

Kind	Distance	Symbol	Plane
axial	$a/2$	a	(010), (001)
axial	$b/2$	b	(001), (100)
axial	$c/2$	c	(010), (100)

Diagonal glide			Diamond glide		
Vector	Symbol	Plane	Vector	Symbol	Plane
$\frac{1}{2}(\mathbf{b}+\mathbf{c})$	n	(100), (0$\bar{1}$1)	$\frac{1}{4}(\mathbf{b}+\mathbf{c})$	d	(100)
$\frac{1}{2}(\mathbf{a}+\mathbf{c})$	n	(010), ($\bar{1}$01)	$\frac{1}{4}(\mathbf{b}+\mathbf{c})$	d	(100)
$\frac{1}{2}(\mathbf{a}+\mathbf{b})$	n	(001), ($\bar{1}$10)	$\frac{1}{4}(\mathbf{a}+\mathbf{c})$	d	(010)
$\frac{1}{2}(\mathbf{a}+\mathbf{b}+\mathbf{c})$	n	(1$\bar{1}$0), (0$\bar{1}$1),	$\frac{1}{4}(\mathbf{a}+\mathbf{b})$	d	(001)
		($\bar{1}$01), (1$\bar{1}$0)	$\frac{1}{4}(\mathbf{a}+\mathbf{b}\pm\mathbf{c})$	d	(1$\bar{1}$0)
$\frac{1}{2}(-\mathbf{a}+\mathbf{b}+\mathbf{c})$	n	(110), (0$\bar{1}$1), ($\bar{1}$01)	$\frac{1}{4}(\pm\mathbf{a}+\mathbf{b}+\mathbf{c})$	d	(01$\bar{1}$)
			$\frac{1}{4}(\mathbf{a}\pm\mathbf{b}+\mathbf{c})$	d	($\bar{1}$01)
$\frac{1}{2}(\mathbf{a}-\mathbf{b}+\mathbf{c})$	n	(011), ($\bar{1}$01), (1$\bar{1}$1)	$\frac{1}{4}(\mathbf{a}\pm\mathbf{b}+\mathbf{c})$	d	($\bar{1}$01)
$\frac{1}{2}(\mathbf{a}+\mathbf{b}-\mathbf{c})$	n	(011), (10$\bar{1}$), ($\bar{1}$10)	$\frac{1}{4}(-\mathbf{a}+\mathbf{b}\pm\mathbf{c})$	d	(110)
			$\frac{1}{4}(\pm\mathbf{a}-\mathbf{b}+\mathbf{c})$	d	(011)
			$\frac{1}{4}(\mathbf{a}\pm\mathbf{b}-\mathbf{c})$	d	(101)

Table 9.3. Definitions of the Strukturbericht Symbols.

Symbol	Definition	Symbol	Definition
A	element	E-K	more complex types
B	AB compounds	L	alloys
C	AB_2	O	organic
D	A_mB_n	S	silicates

Table 9.4. Definitions of the Pearson Symbols.

System:		Centering:	
a	triclinic		
c	cubic	P	primitive
h	hcp	F	face centered
t	tetragonal	I	body centered
m	monoclinic	C	end centered
o	orthorhombic	R	rhombohedral or trigonal

There is no system symbol for trigonal.

Table 9.5. The 32 Point Groups and Their Symbols.

The 5 proper rotations (#1-5):   1,2,3,4,6.

The 5 improper rotations (#6-10): $\bar{1}$, $\bar{2}$ = m, $\bar{3}$, $\bar{4}$, $\bar{6}$ = 3/m.

Combined with with a horizontal mirror (#11- 13):
   $1/m = m = \bar{2}$,   2/m, 3/m = $\bar{6}$,   4/m, 6/m

Combined with a vertical mirror (#14-17):
   2mm,   3mm = 3m,   4mm,   6mm.

Combined with 3 proper rotations (#18-23):
   222,   322 = 32,   422,   622,   23,   432.

Combined with two $\bar{2}$ axes with a rotation axis (vertical and horizontal mirror intersection is a rotation axis) (#24-29):
   2/m2/m2/m = mmm,   3/m m2 = $\bar{6}$m2,   4/m 2/m 2/m = 4/m mm,
   6/m 2/m 2/m = 6/m mm,   2/m $\bar{3}$,   4/m $\bar{3}$ 2/m = m$\bar{3}$m.

Combined with $\bar{3}$ and $\bar{4}$ with proper rotation axes (# 30-32):
   $\bar{4}$2m, $\bar{3}$2/m, $\bar{4}$3m

Table 9.6. Various Symbols used for Point Groups.

Crystal System	International		Schoenflies Symbol	Shubnikov Symbol	Symmetry Formula	Name of Class
	Full	Abbreviated				
Triclinic	1	1	C_1	1	$L1$	monohedral
	$\bar{1}$	$\bar{1}$	C_i (S_2)	$\tilde{2}$	C	pinacoidal
Monoclinic	m	m	C_{1h}, C_s	m	P	dihedral axisless
	2	2	C_2	2	L^2	dihedral axial
	2/m	2/m	C_{2h}	2:m	L^2PC	prismatic
Orthorhombic	2mm	mm	C_{2v}	2·m	L^22P	rhombopyramidal
	222	222	D_2, V	2:2	$3L2$	rhombotetrahedral
	2/m2/m2/m	mmm	D_{2h}, V_h	m·2:m	$3L^23PC$	rhombodipyramidal
Tetragonal:	$\bar{4}$	$\bar{4}$	S_4	$\tilde{4}$	L^24	tetragonal-tetrahedral
	4	4	C_4	4	L^4	tetragonal-pyramidal
	4/m	4/m	C_{4h}	4:m	L^4PC	tetragonal-dipyramidal
	$\bar{4}2m$	$\bar{4}2m$	D_{2d}, V_d	$\tilde{4}$·m	L^242L^22P	tetragonal-scalenohedral
	4mm	4mm	C_{4v}	4·m	L^44P	ditetragonal-pyramidal
	422	42	D_4	4:2	L^44L^2	tetragonal-trapezohedral
	4/m2/m2/m	4/mmm	D_{4h}	m·4:m	L^44L^25PC	ditetragonal-dipyramidal

Table 9.6. Continued

Crystal System	International — Full	International — Abbreviated	Schoenflies Symbol	Shubnikov Symbol	Symmetry Formula	Name of class
Trigonal:	3	3	C_3	3	L^3	trigonal-pyramidal
	$\bar{3}$	$\bar{3}$	C_{3i}; S_6	$\tilde{6}$	$\tilde{L}^6C$	rhombohedral
	3m	3m	C_{3v}	3·m	L^33P	ditrigonal-pyramidal
	32	32	D_3	3:2	L^33L^2	trigonal-trapezohedral
	$\bar{3}2/m$	$\bar{3}m$	D_{3d}	$\tilde{6}$·m	$\tilde{L}^63L^23PC$	ditrigonal-scalenohedral
Hexagonal:	$\bar{6}$	$\bar{6}$	C_{3h}	3:m	L^3P	trigonal-dipyramidal
	6	6	C_6	6	L^6	hexagonal-pyramidal
	6/m	6/m	C_{6h}	6:m	L^6PC	hexagonal-dipyramidal
	$\bar{6}2m$	$\bar{6}2m$	D_{3h}	m·3:m	L^33L^24P	ditrigonal-dipyramidal
	6mm	6mm	C_{6v}	6·m	L^66P	dihexagonal-pyramidal
	622	62	D_6	6:2	L^66L^2	hexagonal-trapezohedral
	6/m2/m2/m	6/mmm	D_{6h}	m·6:m	L^66L^27PC	dihexagonal-dipyramidal
Cubic:	23	23	T	3/2	$3L^24L^3$	tritetrahedral
	$2/m\bar{3}$	$m\bar{3}$	T_h	?	$3L^24\tilde{L}^63PC$	didodecahedral
	$\bar{4}3m$	$\bar{4}3m$	T_d	$3/\tilde{4}$	$3\tilde{L}^44L^36P$	hexatetrahedral
	432	43	O	3/4	$3L^44L^36L^2$	trioctahedral
	$4/m\bar{3}2/m$	$m\bar{3}m$	O_h	$\tilde{6}/4$	$3L^44\tilde{L}^66L^29PC$	hexoctahedral

Data compiled from Barrett and Massalski [1980], Brown and Forsyth [1979], and Vainstein [1979].

Table 9.7. Strukturbericht Symbols, Archtypes, Pearson Symbol, and Space Group Arranged by Strukturbericht Symbol.

SS	Arch.	PS	SG	SS	Arch.	PS	SG
A1	Cu	cF4	Fm3m	B18	CuS	hP12	P6$_3$/mmc
A2	W	cI2	Im3m	B19	β'AuCd	oP4	Pmma
A3	Mg	hP2	P6$_3$/mmc	B20	FeSi	cP8	P2$_1$3
A3'	αLa	hP4	P6$_3$/mmc	B27	BFe	oP8	Pnma
A4	C	cF8	Fd3m	B31	MnP	oP8	Pnma
A5	Sn	tF4	I4$_1$/amd	B32	NaTl	cF16	Fd3m
A6	In	tI2	I4/mmm	B33	xsi-CrB	oC8	Cmcm
A7	As	hR2	R3m	B34	PdS	tP16	P4$_2$/m
A8	Se	hP3	P3$_1$21	B35	CoSn	hP6	P6/mmm
A9	C(graphite)	hP4	P6$_3$/mmc	B37	SeTl	tI16	I4/mcm
A10	Hg	hR1	R3m	B8$_1$	AsNi	hP4	P6$_3$/mmc
A11	Ga	oC8	Cmca	B8$_2$	InNi$_2$	hP4	P6$_3$/mmc
A12	α-Mn	cI58	I43m	B$_a$	CoU	cI16	I2$_1$3
A13	β-Mn	cP20	P4$_1$32	B$_b$	zAgZn	hP9	P3
A15	W$_3$O	cP8	Pm3n	B$_e$	CdSb	oP16	Pbca
A16	αS	oF128	Fddd	B$_f$	xsi-CrB	oC8	Cmcm
A20	αU	oC4	Cmcm	B$_g$	MoB	tI16	I4$_1$/amd
Aa	αPa	tI2	I4/mmm	B$_h$	WC	hP2	P6m2
Ab	βU	tP30	P4$_2$/mnm	B$_i$	γ'-CMo(AsTi)	hP8	P6$_3$/mmc
Ac	αNp	oP8	Pnma	Bl	AsS	mP32	P2$_1$/c
Ad	βNp	tP4	P42$_1$2				
Af	HgSn$_{6-10}$	hP1	P6/mmm	C$_a$	Mg$_2$Ni	hP18	P6$_2$22
Ag	γB	tP50	P4$_2$/nnm	C$_b$	CuMg$_2$	oF48	Fddd
Ah	αPo	cP1	Pm3m	C$_c$	Si$_2$Th	tI12	I4$_1$/amd
				C$_e$	CoGe$_2$	oC23	Aba2
B1	NaCl	cF8	Fm3m	C$_g$	ThC$_2$	mC12	C2/c
B2	CsCl	cP2	Pm3m	Ch	Cu$_2$Te	hP6	P6/mmm
B3	ZnS	cF8	F43m	C1	CaF$_2$	cF12	Fm3m
B4	ZnS	hP4	P6$_3$mc	C1$_b$	AgAsMg	cF12	F43m
B9	HgS	hP6	P3$_1$21	C2	FeS$_2$	cP12	Pa3
B10	PbO	tP4	P4/nmm	C3	Cu$_2$O	cP6	Pn3m
B11	g-CuTi	tP4	P4/nmm	C4	TiO$_2$	tP6	P4$_2$/mnm
B13	α-NiS	hR6	R3m	C6	CdI$_2$	hP3	P3m1
B16	GeS	oP8	Pnma	C7	MoS$_2$	hP6	P6$_3$/mmc
B17	PtS	tP4	P4$_2$/mmc	C11$_a$	CaC$_2$	tI6	I4/mmm

Table 9.7. Continued.

SS	Arch.	PS	SG	SS	Arch.	PS	SG
C11$_b$	MoSi$_2$	tI6	I4/mmm	D0$_{24}$	Ni$_3$Ti	hP16	P6$_3$/mmc
C12	CaSi$_2$	hR6	R$\bar{3}$m	D1$_a$	MoNi$_4$	tI10	I4/m
C14	MgZn$_2$	hP12	P6$_3$/mmc	D1$_b$	Al$_4$U	oI20	Imma
C15	Cu$_2$Mg	cF24	Fd3m	D1$_c$	PtSn$_4$	oC20	Aba2
C15b	AuBe$_5$	cF24	F$\bar{4}$3m	D1$_d$	Pb$_4$Pt	tP10	P4/nbm
C16	Al$_2$Cu	tI12	I4/mcm	D1$_e$	B$_4$Th	tP20	P4/mbm
C18	FeS$_2$	oP6	Pnnm	D1$_f$	BMn$_4$	oF40	Fddd
C19	CdCl$_2$	hR3	R$\bar{3}$m	D1$_g$	B$_4$C	hR15	R$\bar{3}$m
C22	Fe$_2$P	hP9	P$\bar{6}$2m	D13	Al$_4$Ba	tI10	I4/mmm
C23	PbCl$_2$	oP12	Pnma	D2$_b$	Mn$_{12}$Th	tI26	I4/mmm
C32	AlB$_2$	hP3	P6/mmm	D2$_c$	MnU$_6$	tI28	I4/mcm
C33	Bi$_2$STe$_2$	hR5	R$\bar{3}$m	D2$_d$	CaCu$_5$	hP6	P6/mmm
C34	AuTe$_2$	mC6	C2m	D2$_e$	BaHg$_{11}$	cP36	Pm3m
C36	MgNi$_2$	hP24	P6$_3$/mmc	D2$_f$	UB$_{12}$	cF52	Fm3m
C38	Cu$_2$Sb	tP6	P4/nmm	D2$_h$	Al$_6$Mn	oC28	Cmcm
C40	CrSi$_2$	hP9	P6$_2$22	D2$_1$	B$_6$Ca	cP7	Pm3m
C42	SiS$_2$	oI12	Ibam	D2$_3$	NaZn$_{13}$	cF113	Fm3c
C44	GeS$_2$	oF72	Fdd2	D5$_a$	Si$_2$U$_3$	tP10	P4/mbm
C46	AuTe$_2$	oP24	Pma2	D5$_b$	Pt$_2$Sn$_3$	hP10	P6$_3$/mmc
C49	Si$_2$Zr	oC12	Cmcm	D5$_c$	C$_3$Pu$_2$	cI40	I$\bar{4}$3d
C54	Si$_2$Ti	oF24	Fddd	D5$_f$	As$_2$S$_3$	mP20	P2$_1$/c
				D5$_1$	α-Al$_2$O$_3$	hR10	R$\bar{3}$c
D0$_a$	βCu$_3$Ti	oP8	Pmmn	D5$_2$	La$_2$O$_3$	hP5	P$\bar{3}$m1
D0$_c$	SiU$_3$	tI16	I4/mcm	D5$_3$	Mn$_2$O$_3$	cI80	Ia3
D0$_d$	AsMn$_3$	oP16	Pmmn	D5$_8$	S$_3$Sb$_2$	oP20	Pnma
D0$_e$	Ni$_3$P	tI32	I$\bar{4}$	D5$_9$	P$_2$Zn$_3$	tP40	P4$_2$/nmc
D0$_2$	As$_3$Co	cI32	Im3	D5$_{10}$	Cr$_3$C$_2$	oP20	Pnma
D0$_3$	BiF$_3$	cF16	Fm3m	D5$_{13}$	Al$_3$Ni$_2$	hP5	P$\bar{3}$m1
D0$_9$	O$_3$Re	cP4	Pm3m	D7$_a$	δNi$_3$Sn$_4$	mC14	C2/m
D0$_{11}$	Fe$_3$C	oP16	P6/mmm	D7$_b$	Ta$_3$B$_4$	oI14	Immm
D0$_{18}$	AsNa$_3$	hP8	P6$_3$/mmc	D7$_1$	Al$_4$C$_3$	hR7	R$\bar{3}$m
D0$_{19}$	Ni$_3$Sn	hP8	P6$_3$/mmc	D7$_3$	P$_4$Th$_3$	cI28	I$\bar{4}$3d
D0$_{20}$	Al$_3$Ni	oP16	Pnma	D8$_a$	Mn$_{23}$Th$_6$	cF116	Fm3m
D0$_{21}$	Cu$_3$P	hP24	P6$_3$/cm	D8$_b$	σ-CrFe	tP30	P4$_2$/mnm
D0$_{22}$	Al$_3$Ti	tI8	I4/mmm	D8$_c$	Mg$_2$Zn$_{11}$	cP39	Pm3
D0$_{23}$	Al$_3$Zr	tI16	I4/mmm	D8$_d$	Co$_2$Al$_9$	mP22	P2$_1$/c

Table 9.7. Continued.

SS	Arch.	PS	SG	SS	Arch.	PS	SG
$D8_e$	$Mg_{32}(Al,Zn)_{49}$	cI162	Im3	$F0_1$	NiSSb	cP12	$P2_13$
$D8_f$	Ge_7Ir_3	cI40	Im3m	$F5_a$	$FeKS_2$	mC16	C2/c
$D8_g$	Ga_2Mg_5	oI28	Ibam	$F5_1$	$CrNaS_2$	hR4	R3m
$D8_h$	B_5W_2	hP14	P6$_3$/mmc	$F5_6$	CuS_2Sb	oP16	Pnma
$D8_i$	Mo_2B_5	hR7	R3m				
$D8_k$	Th_7S_{12}	hP20	P6$_3$/m	$H1_1$	Al_2MgO_4	cF56	Fd3m
$D8_l$	Cr_5Br_3	tI32	I4/mcm	$H2_4$	Cu_3S_4V	cP8	P43m
$D8_m$	Si_3W_5	tI32	I4/mcm				
$D8_1$	Fe_3Zn_{10}	cI52	Im3m	$L'2_b$	H_2Th	tI6	I4/mmm
$D8_2$	Cu_3Zn_8	cI52	I43m	L'3	Fe_2N	hP3	P6$_3$/mmc
$D8_3$	Al_4Cu_9	cP52	P43m	$L1_0$	AuCu(I)	tP4	P4/mmm
$D8_4$	$Cr_{23}C_6$	cF116	Fm3m	$L1_2$	$AuCu_3$	cP4	Pm3m
$D8_5$	Fe_7W_6	hR13	R3m	$L2_1$	$AlCu_2Mn$	cF16	Fm3m
$D8_6$	$Cu_{15}Si_4$	cI76	I43d	$L2_2$	Sb_2Tl_7	cI54	Im3m
$D8_8$	Mn_5Si_3	hP16	P6$_3$/mcm	$L6_0$	$CuTi_3$	tP4	P4/mmm
$D8_9$	Co_9S_8	cF68	Fm3m				
$D8_{10}$	Al_8Cr_5	hR26	R3m				
$D8_{11}$	Al_5Co_2	hP28	P6$_3$/mmc				
$D10_1$	Cr_7C_3	hP80	Pnma				
$D10_2$	Fe_3Th_7	hP20	P6$_3$mc				
$E0_1$	ClFPb	tP6	P4/nmm				
$E1_a$	BRe_3	oC16	Cmcm				
$E1_b$	$AgAuTe_4$	mP12	P2/c				
$E1_1$	$CuFeS_2$	tI16	I42d				
$E2_1$	CaO_3Ti	cP5	Pm3m				
E3	Al_2CdS_4	tI14	I4				
$E9_a$	Al_7Cu_2Fe	tP40	P4/mnc				
$E9_b$	$AlLi_3N_2$	cI96	P62m				
$E9_d$	$AlLi_3N_2$	cI96	Ia3				
$E9_e$	$CuFe_2S_3$	oP24	Pnma				
$E9_3$	Fe_3W_3C	cF11	Fd3m				
$E9_4$	Al_4C_4Si	hP18	P6$_3$mc				

[After T. B. Massalski (Editor), *Binary Alloy Phase Diagrams*, **2**, ASMI, Metals Park, 1986, p. 2170; by permission.]

Table 9.8. Strukturbericht Symbols, Archtypes, Pearson Symbol, and Space Group Arranged by Archtype.

SS	Arch.	PS	SG	SS	Arch.	PS	SG
$D5_1$	α-Al$_2$O$_3$	hR10	$R\bar{3}c$	Bl	AsS	mP32	P2$_1$/c
A12	α-Mn	cI58	$I\bar{4}3m$	A20	αU	oC4	Cmcm
B13	α-NiS	hR6	R3m	C15b	AuBe$_5$	cF24	$F\bar{4}3m$
C1$_b$	AgAsMg	cF12	$F\bar{4}3m$	L1$_0$	AuCu(I)	tP4	P4/mmm
E1$_b$	AgAuTe$_4$	mP12	P2/c	L1$_2$	AuCu$_3$	cP4	Pm3m
E3	Al$_2$CdS$_4$	tI14	$I\bar{4}$	C34	AuTe$_2$	mC6	C2m
C16	Al$_2$Cu	tI12	I4/mcm	C46	AuTe$_2$	oP24	Pma2
H1$_1$	Al$_2$MgO$_4$	cF56	Fd3m				
D0$_{20}$	Al$_3$Ni	oP16	Pnma	B19	β'AuCd	oP4	Pmma
D5$_{13}$	Al$_3$Ni$_2$	hP5	$P\bar{3}m1$	A13	β-Mn	cP20	P4$_1$32
D0$_{22}$	Al$_3$Ti	tI8	I4/mmm	D1$_g$	B$_4$C	hR15	R3m
D0$_{23}$	Al$_3$Zr	tI16	I4/mmm	D1$_e$	B$_4$Th	tP20	P4/mbm
D1$_3$	Al$_4$Ba	tI10	I4/mmm	D8$_h$	B$_5$W$_2$	hP14	P6$_3$/mmc
D7$_1$	Al$_4$C$_3$	hR7	R3m	D2$_1$	B$_6$Ca	cP7	Pm3m
E9$_4$	Al$_4$C$_4$Si	hP18	P6$_3$mc	D2$_e$	BaHg$_{11}$	cP36	Pm3m
D8$_3$	Al$_4$Cu$_9$	cP52	$P\bar{4}3m$	D0$_a$	βCu$_3$Ti	oP8	Pmmn
D1$_b$	Al$_4$U	oI20	Imma	B27	BFe	oP8	Pnma
D8$_{11}$	Al$_5$Co$_2$	hP28	P6$_3$/mmc	C33	Bi$_2$STe$_2$	hR5	R3m
D2$_h$	Al$_6$Mn	oC28	Cmcm	D0$_3$	BiF$_3$	cF16	Fm3m
E9$_a$	Al$_7$Cu$_2$Fe	tP40	P4/mnc	D1$_f$	BMn$_4$	oF40	Fddd
D8$_{10}$	Al$_8$Cr$_5$	hR26	R3m	Ad	βNp	tP4	P42$_1$2
A3'	αLa	hP4	P6$_3$/mmc	E1$_a$	BRe$_3$	oC16	Cmcm
C32	AlB$_2$	hP3	P6/mmm	Ab	βU	tP30	P42/mnm
L2$_1$	AlCu$_2$Mn	cF16	Fm3m				
E9$_b$	AlLi$_3$N$_2$	cI96	$P\bar{6}2m$	A4	C	cF8	Fd3m
E9$_d$	AlLi$_3$N$_2$	cI96	Ia3	A9	C(graphite)	hP4	P6$_3$/mmc
Ac	αNp	oP8	Pnma	D5$_c$	C$_3$Pu$_2$	cI40	$I\bar{4}3d$
Aa	αPa	tI2	I4/mmm	C11$_a$	CaC$_2$	tI6	I4/mmm
Ah	αPo	cP1	Pm3m	D2$_d$	CaCu$_5$	hP6	P6/mmm
A7	As	hR2	R3m	C1	CaF$_2$	cF12	Fm3m
A16	αS	oF128	Fddd	E2$_1$	CaO$_3$Ti	cP5	Pm3m
D5$_f$	As$_2$S$_3$	mP20	P2$_1$/c	C12	CaSi$_2$	hR6	R3m
D0$_2$	As$_3$Co	cI32	Im3	C19	CdCl$_2$	hR3	R3m
D0$_d$	AsMn$_3$	oP16	Pmmn	C6	CdI$_2$	hP3	P3m1
D0$_{18}$	AsNa$_3$	hP8	P6$_3$/mmc	B$_e$	CdSb	oP16	Pbca
B8$_1$	AsNi	hP4	P6$_3$/mmc	E0$_1$	ClFPb	tP6	P4/nmm

Table 9.8. Continued.

SS	Arch.	PS	SG	SS	Arch.	PS	SG
D8$_d$	Co_2Al_9	mP22	P2$_1$/c	D8$_5$	Fe_7W_6	hR13	R$\bar{3}$m
D8$_9$	Co_9S_8	cF68	Fm3m	F5$_a$	$FeKS_2$	mC16	C2/c
C$_e$	$CoGe_2$	oC23	Aba2	C2	FeS_2	cP12	Pa3
B35	CoSn	hP6	P6/mmm	C18	FeS_2	oP6	Pnnm
B$_a$	CoU	cI16	I2$_1$3	B20	FeSi	cP8	P2$_1$3
D5$_{10}$	Cr_3C_2	oP20	Pnma				
D8$_l$	Cr_5Br_3	tI32	I4/mcm	B$_i$	γ'-CMo(AsTi)	hP8	P6$_3$/mmc
D10$_1$	Cr_7C_3	hP80	Pnma	B11	γ-CuTi	tP4	P4/nmm
D8$_4$	$Cr_{23}C_6$	cF116	Fm3m	A11	Ga	oC8	Cmca
F5$_1$	$CrNaS_2$	hR4	R$\bar{3}$m	D8$_g$	Ga_2Mg_5	oI28	Ibam
C40	$CrSi_2$	hP9	P6$_2$22	Ag	γB	tP50	P4$_2$/nnm
B2	CsCl	cP2	Pm3m	D8$_f$	Ge_7Ir_3	cI40	Im3m
A1	Cu	cF4	Fm3m	B16	GeS	oP8	Pnma
C15	Cu_2Mg	cF24	Fd3m	C44	GeS_2	oF72	Fdd2
C3	Cu_2O	cP6	Pn3m				
C38	Cu_2Sb	tP6	P4/nmm	L'2$_b$	H_2Th	tI6	I4/mmm
Ch	Cu_2Te	hP6	P6/mmm	A10	Hg	hR1	R$\bar{3}$m
D0$_{21}$	Cu_3P	hP24	P6$_3$/cm	B9	HgS	hP6	P3$_1$21
H2$_4$	Cu_3S_4V	cP8	P$\bar{4}$3m	Af	$HgSn_{6-10}$	hP1	P6/mmm
D8$_2$	Cu_3Zn_8	cI52	I$\bar{4}$3m				
D8$_6$	$Cu_{15}Si_4$	cI76	I$\bar{4}$3d	A6	In	tI2	I4/mmm
E9$_e$	$CuFe_2S_3$	oP24	Pnma	B8$_2$	$InNi_2$	hP4	P6$_3$/mmc
E1$_1$	$CuFeS_2$	tI16	I$\bar{4}$2d				
C$_b$	$CuMg_2$	oF48	Fddd	D52	La_2O_3	hP5	P$\bar{3}$m1
B18	CuS	hP12	P6$_3$/mmc				
F5$_6$	CuS_2Sb	oP16	Pnma	A3	Mg	hP2	P6$_3$/mmc
L6$_0$	$CuTi_3$	tP4	P4/mmm	C$_a$	Mg_2Ni	hP18	P6$_2$22
				D8$_c$	Mg_2Zn_{11}	cP39	Pm3
D7$_a$	δNi_3Sn_4	mC14	C2/m	D8$_e$	$Mg_{32}(Al,Zn)_{49}$	cI162	Im3
				C36	$MgNi_2$	hP24	P6$_3$/mmc
L'3	Fe_2N	hP3	P6$_3$/mmc	C14	$MgZn_2$	hP12	P6$_3$/mmc
C22	Fe_2P	hP9	P$\bar{6}$2m	D5$_3$	Mn_2O_3	cI80	Ia3
D0$_{11}$	Fe_3C	oP16	P6/mmm	D8$_8$	Mn_5Si_3	hP16	P6$_3$/mcm
D10$_2$	Fe_3Th_7	hP20	P6$_3$mc	D2$_b$	$Mn_{12}Th$	tI26	I4/mmm
E9$_3$	Fe_3W_3C	cF11	Fd3m	D8$_a$	$Mn_{23}Th_6$	cF116	Fm3m
D8$_1$	Fe_3Zn_{10}	cI52	Im3m	B31	MnP	oP8	Pnma

Table 9.8. Continued.

SS	Arch.	PS	SG	SS	Arch.	PS	SG
$D2_c$	MnU_6	tI28	I4/mcm	C42	SiS_2	oI12	Ibam
B_g	MoB	tI16	$I4_1$/amd	$D0_c$	SiU_3	tI16	I4/mcm
$D1_a$	$MoNi_4$	tI10	I4/m	A5	Sn	tF4	$I4_1$/amd
$D8_i$	Mo_2B_5	hR7	R3m	$D7_b$	Ta_3B_4	oI14	Immm
C7	MoS_2	hP6	$P6_3$/mmc	$D8_k$	Th_7S_{12}	hP20	$P6_3$/m
$C11_b$	$MoSi_2$	tI6	I4/mmm	C_g	ThC_2	mC12	C2/c
				C4	TiO_2	tP6	$P4_2$/mnm
B1	NaCl	cF8	Fm3m				
B32	NaTl	cF16	Fd3m	$D2_f$	UB_{12}	cF52	Fm3m
$D2_3$	$NaZn_{13}$	cF113	Fm3c	A2	W	cI2	Im3m
$D0_e$	Ni_3P	tI32	$I\bar{4}$	A15	W_3O	cP8	Pm3n
$D0_{19}$	Ni_3Sn	hP8	$P6_3$/mmc	B_h	WC	hP2	$P\bar{6}m2$
$D0_{24}$	Ni_3Ti	hP16	$P6_3$/mmc				
$F0_1$	NiSSb	cP12	$P2_13$	B_f	xsi-CrB	oC8	Cmcm
				B33	xsi-CrB	oC8	Cmcm
$D0_9$	O_3Re	cP4	Pm3m				
				B_b	ζAgZn	hP9	P3
$D5_9$	P_2Zn_3	tP40	$P4_2$/nmc	B3	ZnS	cF8	$F\bar{4}3m$
$D7_3$	P_4Th_3	cI28	$I\bar{4}3d$	B4	ZnS	hP4	$P6_3mc$
$D1_d$	Pb_4Pt	tP10	P4/nbm				
C23	$PbCl_2$	oP12	Pnma				
B10	PbO	tP4	P4/nmm				
B34	PdS	tP16	$P4_2$/m				
$D5_b$	Pt_2Sn_3	hP10	$P6_3$/mmc				
B17	PtS	tP4	$P4_2$/mmc				
$D1_c$	$PtSn_4$	oC20	Aba2				
$D8_b$	σ-CrFe	tP30	$P4_2$/mnm				
$D5_8$	S_3Sb_2	oP20	Pnma				
$L2_2$	Sb_2Tl_7	cI54	Im3m				
A8	Se	hP3	$P3_121$				
B37	SeTl	tI16	I4/mcm				
C_c	Si_2Th	tI12	$I4_1$/amd				
C54	Si_2Ti	oF24	Fddd				
$D5_a$	Si_2U_3	tP10	P4/mbm				
C49	Si_2Zr	oC12	Cmcm				
$D8_m$	Si_3W_5	tI32	I4/mcm				

[After T. B. Massalski (Editor), *Binary Alloy Phase Diagrams*, **2**, ASMI, Metals Park, 1986, p. 2170; by permission.]

Table 9.9. Strukturbericht Symbols, Archtypes, Pearson Symbol, and Space Group Arranged by Pearson Symbol.

SS	Arch.	PS	SG	SS	Arch.	PS	SG
A1	Cu	cF4	Fm3m	$D0_9$	O_3Re	cP4	Pm3m
A4	C	cF8	Fd3m	$E2_1$	CaO_3Ti	cP5	Pm3m
B1	NaCl	cF8	Fm3m	C3	Cu_2O	cP6	Pn3m
B3	ZnS	cF8	F43m	$D2_1$	B_6Ca	cP7	Pm3m
$E9_3$	Fe_3W_3C	cF11	Fd3m	$H2_4$	Cu_3S_4V	cP8	P43m
$C1_b$	AgAsMg	cF12	F43m				
C1	CaF_2	cF12	Fm3m	B20	FeSi	cP8	$P2_13$
$L2_1$	$AlCu_2Mn$	cF16	Fm3m	A15	W_3O	cP8	Pm3n
$D0_3$	BiF_3	cF16	Fm3m	C2	FeS_2	cP12	Pa3
B32	NaTl	cF16	Fd3m	$F0_1$	NiSSb	cP12	$P2_13$
$C15_b$	$AuBe_5$	cF24	F43m	A13	β-Mn	cP20	$P4_132$
C15	Cu_2Mg	cF24	Fd3m	$D2_e$	$BaHg_{11}$	cP36	Pm3m
$D2_f$	UB_{12}	cF52	Fm3m	$D8_c$	Mg_2Zn_{11}	cP39	Pm3
$H1_1$	Al_2MgO_4	cF56	Fd3m	$D8_3$	Al_4Cu_9	cP52	P43m
$D8_9$	Co_9S_8	cF68	Fm3m				
$D2_3$	$NaZn_{13}$	cF113	Fm3c	Af	$HgSn_{6-10}$	hP1	P6/mmm
$D8_4$	$Cr_{23}C_6$	cF116	Fm3m	A3	Mg	hP2	$P6_3/mmc$
$D8_a$	$Mn_{23}Th_6$	cF116	Fm3m	B_h	WC	hP2	P6m2
A2	W	cI2	Im3m	C32	AlB_2	hP3	P6/mmm
B_a	CoU	cI16	$I2_13$	C6	CdI_2	hP3	P3m1
$D7_3$	P_4Th_3	cI28	I43d	L'3	Fe_2N	hP3	$P6_3/mmc$
$D0_2$	As_3Co	cI32	Im3	A8	Se	hP3	$P3_121$
$D5_c$	C_3Pu_2	cI40	I43d	A3'	αLa	hP4	$P6_3/mmc$
$D8_f$	Ge_7Ir_3	cI40	Im3m	$B8_1$	AsNi	hP4	$P6_3/mmc$
$D8_2$	Cu_3Zn_8	cI52	I43m	A9	C(graphite)	hP4	$P6_3/mmc$
$D8_1$	Fe_3Zn_{10}	cI52	Im3m	$B8_2$	$InNi_2$	hP4	$P6_3/mmc$
$L2_2$	Sb_2Tl_7	cI54	Im3m	B4	ZnS	hP4	$P6_3mc$
A12	α-Mn	cI58	I43m	$D5_{13}$	Al_3Ni_2	hP5	P3m1
$D8_6$	$Cu_{15}Si_4$	cI76	I43d	D52	La_2O_3	hP5	P3m1
$D5_3$	Mn_2O_3	cI80	Ia3	$D2_d$	$CaCu_5$	hP6	P6/mmm
$E9_b$	$AlLi_3N_2$	cI96	P62m	B35	CoSn	hP6	P6/mmm
$E9_d$	$AlLi_3N_2$	cI96	Ia3	Ch	Cu_2Te	hP6	P6/mmm
$D8_e$	$Mg_{32}(Al,Zn)_{49}$	cI162	Im3	B9	HgS	hP6	$P3_121$
Ah	αPo	cP1	Pm3m	C7	MoS_2	hP6	$P6_3/mmc$
B2	CsCl	cP2	Pm3m	$D0_{18}$	$AsNa_3$	hP8	$P6_3/mmc$
$L1_2$	$AuCu_3$	cP4	Pm3m	B_i	γ'-CMo(AsTi)	hP8	$P6_3/mmc$

Table 9.9. Continued.

SS	Arch.	PS	SG	SS	Arch.	PS	SG
$D0_{19}$	Ni_3Sn	hP8	$P6_3/mmc$	$E1_b$	$AgAuTe_4$	mP12	P2/c
C40	$CrSi_2$	hP9	$P6_222$	$D5_f$	As_2S_3	mP20	$P2_1/c$
C22	Fe_2P	hP9	$P\bar{6}2m$	$D8_d$	Co_2Al_9	mP22	$P2_1/c$
B_b	$\zeta AgZn$	hP9	$P\bar{3}$	B1	AsS	mP32	$P2_1/c$
$D5_b$	Pt_2Sn_3	hP10	$P6_3/mmc$				
B18	CuS	hP12	$P6_3/mmc$	A20	αU	oC4	Cmcm
C14	$MgZn_2$	hP12	$P6_3/mmc$	A11	Ga	oC8	Cmca
$D8_h$	B_5W_2	hP14	$P6_3/mmc$	B_f	xsi-CrB	oC8	Cmcm
$D8_g$	Mn_5Si_3	hP16	$P6_3/mcm$	B33	xsi-CrB	oC8	Cmcm
$D0_{24}$	Ni_3Ti	hP16	$P6_3/mmc$	C49	Si_2Zr	oC12	Cmcm
$E9_4$	Al_4C_4Si	hP18	$P6_3mc$	$E1_a$	BRe_3	oC16	Cmcm
C_a	Mg_2Ni	hP18	$P6_222$	$D1_c$	$PtSn_4$	oC20	Aba2
$D10_2$	Fe_3Th_7	hP20	$P6_3mc$	C_e	$CoGe_2$	oC23	Aba2
$D8_k$	Th_7S_{12}	hP20	$P6_3/m$	$D2_h$	Al_6Mn	oC28	Cmcm
$D0_{21}$	Cu_3P	hP24	$P6_3/cm$	C54	Si_2Ti	oF24	Fddd
C36	$MgNi_2$	hP24	$P6_3/mmc$	$D1_f$	BMn_4	oF40	Fddd
$D8_{11}$	Al_5Co_2	hP28	$P6_3/mmc$	C_b	$CuMg_2$	oF48	Fddd
$D10_1$	Cr_7C_3	hP80	Pnma	C44	GeS_2	oF72	Fdd2
A10	Hg	hR1	$R\bar{3}m$	A16	αS	oF128	Fddd
A7	As	hR2	$R\bar{3}m$	C42	SiS_2	oI12	Ibam
C19	$CdCl_2$	hR3	$R\bar{3}m$	$D7_b$	Ta_3B_4	oI14	Immm
$F5_1$	$CrNaS_2$	hR4	$R\bar{3}m$	$D1_b$	Al_4U	oI20	Imma
C33	Bi_2STe_2	hR5	$R\bar{3}m$	$D8_g$	Ga_2Mg_5	oI28	Ibam
B13	α-NiS	hR6	R3m	B19	$\beta'AuCd$	oP4	Pmma
C12	$CaSi_2$	hR6	$R\bar{3}m$	C18	FeS_2	oP6	Pnnm
$D7_1$	Al_4C_3	hR7	$R\bar{3}m$	Ac	αNp	oP8	Pnma
$D8_i$	Mo_2B_5	hR7	$R\bar{3}m$	$D0_a$	βCu_3Ti	oP8	Pmmn
$D5_1$	α-Al_2O_3	hR10	$R\bar{3}c$	B27	BFe	oP8	Pnma
$D8_5$	Fe_7W_6	hR13	$R\bar{3}m$	B16	GeS	oP8	Pnma
$D1_g$	B_4C	hR15	$R\bar{3}m$	B31	MnP	oP8	Pnma
$D8_{10}$	Al_8Cr_5	hR26	R3m	C23	$PbCl_2$	oP12	Pnma
				$D0_{20}$	Al_3Ni	oP16	Pnma
C34	$AuTe_2$	mC6	C2m	$D0_d$	$AsMn_3$	oP16	Pmmn
C_g	ThC_2	mC12	C2/c	B_e	CdSb	oP16	Pbca
$D7_a$	dNi_3Sn_4	mC14	C2/m	$F5_6$	CuS_2Sb	oP16	Pnma
$F5_a$	$FeKS_2$	mC16	C2/c	$D0_{11}$	Fe_3C	oP16	P6/mmm

Table **9.9.** Continued.

SS	Arch.	PS	SG	SS	Arch.	PS	SG
$D5_{10}$	Cr_3C_2	oP20	Pnma	C38	Cu_2Sb	tP6	P4/nmm
$D5_8$	S_3Sb_2	oP20	Pnma	C4	TiO_2	tP6	$P4_2$/mnm
C46	$AuTe_2$	oP24	Pma2	$D1_d$	Pb_4Pt	tP10	P4/nbm
$E9_e$	$CuFe_2S_3$	oP24	Pnma	$D5_a$	Si_2U_3	tP10	P4/mbm
				B34	PdS	tP16	$P4_2$/m
A5	Sn	tF4	$I4_1$/amd	$D1_e$	B_4Th	tP20	P4/mbm
Aa	αPa	tI2	I4/mmm	Ab	βU	tP30	$P4_2$/mnm
A6	In	tI2	I4/mmm	$D8_b$	σ-CrFe	tP30	$P4_2$/mnm
$C11_a$	CaC_2	tI6	I4/mmm	$E9_a$	Al_7Cu_2Fe	tP40	P4/mnc
$L'2_b$	H_2Th	tI6	I4/mmm	D59	P_2Zn_3	tP40	$P4_2$/nmc
$C11_b$	$MoSi_2$	tI6	I4/mmm	Ag	γB	tP50	$P4_2$/nnm
$D0_{22}$	Al_3Ti	tI8	I4/mmm				
$D1_3$	Al_4Ba	tI10	I4/mmm				
$D1_a$	$MoNi_4$	tI10	I4/m				
C16	Al_2Cu	tI12	I4/mcm				
C_c	Si_2Th	tI12	$I4_1$/amd				
E3	Al_2CdS_4	tI14	$I\bar{4}$				
$D0_{23}$	Al_3Zr	tI16	I4/mmm				
$E1_1$	$CuFeS_2$	tI16	$I\bar{4}2d$				
B_g	MoB	tI16	$I4_1$/amd				
B37	SeTl	tI16	I4/mcm				
$D0_c$	SiU_3	tI16	I4/mcm				
$D2_b$	$Mn_{12}Th$	tI26	I4/mmm				
$D2_c$	MnU_6	tI28	I4/mcm				
$D8_1$	Cr_5Br_3	tI32	I4/mcm				
$D0_e$	Ni_3P	tI32	$I\bar{4}$				
$D8_m$	Si_3W_5	tI32	I4/mcm				
$L1_0$	AuCu(I)	tP4	P4/mmm				
Ad	βNp	tP4	$P42_12$				
$L6_0$	$CuTi_3$	tP4	P4/mmm				
B11	γ-CuTi	tP4	P4/nmm				
B10	PbO	tP4	P4/nmm				
B17	PtS	tP4	$P4_2$/mmc				
$E0_1$	ClFPb	tP6	P4/nmm				

[After T. B. Massalski (Editor), *Binary Alloy Phase Diagrams*, **2**, ASMI, Metals Park, 1986, p. 2170; by permission.]

Table 9.10. Strukturbericht Symbols, Archtypes, Pearson Symbol, and Space Group Arranged by Space Group.

SS	Arch.	PS	SG	SS	Arch.	PS	SG
C_e	$CoGe_2$	oC23	Aba2	$D8_a$	$Mn_{23}Th_6$	cF116	Fm3m
$D1_c$	$PtSn_4$	oC20	Aba2	$L2_1$	$AlCu_2Mn$	cF16	Fm3m
$C_{\tilde{g}}$	ThC_2	mC12	C2/c	$D1_a$	$MoNi_4$	tI10	I4/m
$F5_a$	$FeKS_2$	mC16	C2/c	B37	SeTl	tI16	I4/mcm
$D7_a$	δNi_3Sn_4	mC14	C2/m	C16	Al_2Cu	tI12	I4/mcm
C34	$AuTe_2$	mC6	C2m	$D0_c$	SiU_3	tI16	I4/mcm
A11	Ga	oC8	Cmca	$D2_c$	MnU_6	tI28	I4/mcm
A20	αU	oC4	Cmcm	$D8_l$	Cr_5Br_3	tI32	I4/mcm
B_f	xsi-CrB	oC8	Cmcm	$D8_m$	Si_3W_5	tI32	I4/mcm
B33	xsi-CrB	oC8	Cmcm	A6	In	tI2	I4/mmm
C49	Si_2Zr	oC12	Cmcm	Aa	αPa	tI2	I4/mmm
$D2_h$	Al_6Mn	oC28	Cmcm	$C11_a$	CaC_2	tI6	I4/mmm
$E1_a$	BRe_3	oC16	Cmcm	$C11_b$	$MoSi_2$	tI6	I4/mmm
				$D0_{22}$	Al_3Ti	tI8	I4/mmm
B3	ZnS	cF8	F$\bar{4}$3m	$D0_{23}$	Al_3Zr	tI16	I4/mmm
$C1_b$	AgAsMg	cF12	F$\bar{4}$3m	$D1_3$	Al_4Ba	tI10	I4/mmm
C15b	$AuBe_5$	cF24	F$\bar{4}$3m	$D2_b$	$Mn_{12}Th$	tI26	I4/mmm
A4	C	cF8	Fd3m	$L'2_b$	H_2Th	tI6	I4/mmm
B32	NaTl	cF16	Fd3m	A5	Sn	tF4	I4$_1$/amd
C15	Cu_2Mg	cF24	Fd3m	B_g	MoB	tI16	I4$_1$/amd
E93	Fe_3W_3C	cF11	Fd3m	C_c	Si_2Th	tI12	I4$_1$/amd
$H1_1$	Al_2MgO_4	cF56	Fd3m	B_a	CoU	cI16	I2$_1$3
C44	GeS_2	oF72	Fdd2	$D0_e$	Ni_3P	tI32	I$\bar{4}$
A16	αS	oF128	Fddd	E3	Al_2CdS_4	tI14	I$\bar{4}$
C54	Si_2Ti	oF24	Fddd	$E1_1$	$CuFeS_2$	tI16	I$\bar{4}$2d
C_b	$CuMg_2$	oF48	Fddd	$D5_c$	C_3Pu_2	cI40	I$\bar{4}$3d
$D1_f$	BMn_4	oF40	Fddd	$D7_3$	P_4Th_3	cI28	I$\bar{4}$3d
$D2_3$	$NaZn_{13}$	cF113	Fm3c	$D8_6$	$Cu_{15}Si_4$	cI76	I$\bar{4}$3d
A1	Cu	cF4	Fm3m	A12	α-Mn	cI58	I$\bar{4}$3m
B1	NaCl	cF8	Fm3m	$D8_2$	Cu_3Zn_8	cI52	I$\bar{4}$3m
C1	CaF_2	cF12	Fm3m	$D5_3$	Mn_2O_3	cI80	Ia3
$D0_3$	BiF_3	cF16	Fm3m	$E9_d$	$AlLi_3N_2$	cI96	Ia3
$D2_f$	UB_{12}	cF52	Fm3m	C42	SiS_2	oI12	Ibam
$D8_4$	$Cr_{23}C_6$	cF116	Fm3m	$D8_g$	Ga_2Mg_5	oI28	Ibam
$D8_9$	Co_9S_8	cF68	Fm3m	$D0_2$	As_3Co	cI32	Im3

Table 9.10. Continued.

SS	Arch.	PS	SG	SS	Arch.	PS	SG
$D8_e$	$Mg_{32}(Al,Zn)_{49}$	cI162	Im3	$D8_k$	Th_7S_{12}	hP20	$P6_3/m$
A2	W	cI2	Im3m	$D8_8$	Mn_5Si_3	hP16	$P6_3/mcm$
$D8_1$	Fe_3Zn_{10}	cI52	Im3m	A3	Mg	hP2	$P6_3/mmc$
$D8_f$	Ge_7Ir_3	cI40	Im3m	A3'	αLa	hP4	$P6_3/mmc$
$L2_2$	Sb_2Tl_7	cI54	Im3m	A9	C(graphite)	hP4	$P6_3/mmc$
$D1_b$	Al_4U	oI20	Imma	$B8_1$	AsNi	hP4	$P6_3/mmc$
$D7_b$	Ta_3B_4	oI14	Immm	$B8_2$	$InNi_2$	hP4	$P6_3/mmc$
				B18	CuS	hP12	$P6_3/mmc$
$E1_b$	$AgAuTe_4$	mP12	P2/c	B_i	γ'-CMo(AsTi)	hP8	$P6_3/mmc$
$D1_e$	B_4Th	tP20	P4/mbm	C7	MoS_2	hP6	$P6_3/mmc$
$D5_a$	Si_2U_3	tP10	P4/mbm	C14	$MgZn_2$	hP12	$P6_3/mmc$
$L1_0$	AuCu(I)	tP4	P4/mmm	C36	$MgNi_2$	hP24	$P6_3/mmc$
$L6_0$	$CuTi_3$	tP4	P4/mmm	$D0_{18}$	$AsNa_3$	hP8	$P6_3/mmc$
$E9_a$	Al_7Cu_2Fe	tP40	P4/mnc	$D0_{19}$	Ni_3Sn	hP8	$P6_3/mmc$
$D1_d$	Pb_4Pt	tP10	P4/nbm	$D0_{24}$	Ni_3Ti	hP16	$P6_3/mmc$
B10	PbO	tP4	P4/nmm	$D5_b$	Pt_2Sn_3	hP10	$P6_3/mmc$
B11	g-CuTi	tP4	P4/nmm	$D8_{11}$	Al_5Co_2	hP28	$P6_3/mmc$
C38	Cu_2Sb	tP6	P4/nmm	$D8_h$	B_5W_2	hP14	$P6_3/mmc$
$E0_1$	ClFPb	tP6	P4/nmm	L'3	Fe_2N	hP3	$P6_3/mmc$
Af	$HgSn_{6-10}$	hP1	P6/mmm	B4	ZnS	hP4	$P6_3mc$
B35	CoSn	hP6	P6/mmm	$D10_2$	Fe_3Th_7	hP20	$P6_3mc$
C32	AlB_2	hP3	P6/mmm	$E9_4$	Al_4C_4Si	hP18	$P6_3mc$
Ch	Cu_2Te	hP6	P6/mmm	B20	FeSi	cP8	$P2_13$
$D0_{11}$	Fe_3C	oP16	P6/mmm	$F0_1$	NiSSb	cP12	$P2_13$
$D2_d$	$CaCu_5$	hP6	P6/mmm	A8	Se	hP3	$P3_121$
Bl	AsS	mP32	$P2_1/c$	B9	HgS	hP6	$P3_121$
$D5_f$	As_2S_3	mP20	$P2_1/c$	A13	β-Mn	cP20	$P4_132$
$D8_d$	Co_2Al_9	mP22	$P2_1/c$	Ad	βNp	tP4	$P4_22_12$
B34	PdS	tP16	$P4_2/m$	C40	$CrSi_2$	hP9	$P6_222$
B17	PtS	tP4	$P4_2/mmc$	C_a	Mg_2Ni	hP18	$P6_222$
Ab	βU	tP30	$P4_2/mnm$	B_b	zAgZn	hP9	P3
C4	TiO_2	tP6	$P4_2/mnm$	C6	CdI_2	hP3	$P\bar{3}m1$
$D8_b$	σ-CrFe	tP30	$P4_2/mnm$	$D5_{13}$	Al_3Ni_2	hP5	$P\bar{3}m1$
$D5_9$	P_2Zn_3	tP40	$P4_2/nmc$	D52	La_2O_3	hP5	$P\bar{3}m1$
Ag	γB	tP50	$P4_2/nnm$	$D8_3$	Al_4Cu_9	cP52	$P\bar{4}3m$
$D0_{21}$	Cu_3P	hP24	$P6_3/cm$	$H2_4$	Cu_3S_4V	cP8	$P\bar{4}3m$

Table 9.10. Continued.

SS	Arch.	PS	SG	SS	Arch.	PS	SG
C22	Fe_2P	hP9	P$\bar{6}$2m	D5$_1$	α-Al_2O_3	hR10	R$\bar{3}$c
E9$_b$	$AlLi_3N_2$	cI96	P$\bar{6}$2m	A7	As	hR2	R$\bar{3}$m
B$_h$	WC	hP2	P$\bar{6}$m2	A10	Hg	hR1	R$\bar{3}$m
C2	FeS_2	cP12	Pa3	C12	$CaSi_2$	hR6	R$\bar{3}$m
B$_e^-$	CdSb	oP16	Pbca	C19	$CdCl_2$	hR3	R$\bar{3}$m
D8$_c$	Mg_2Zn_{11}	cP39	Pm3	C33	Bi_2STe_2	hR5	R$\bar{3}$m
Ah	αPo	cP1	Pm3m	D1$_g$	B_4C	hR15	R$\bar{3}$m
B2	CsCl	cP2	Pm3m	D7$_1$	Al_4C_3	hR7	R$\bar{3}$m
D0$_9$	O_3Re	cP4	Pm3m	D8$_5$	Fe_7W_6	hR13	R$\bar{3}$m
D2$_1$	B_6Ca	cP7	Pm3m	D8$_i$	Mo_2B_5	hR7	R$\bar{3}$m
D2$_e$	$BaHg_{11}$	cP36	Pm3m	F5$_1$	$CrNaS_2$	hR4	R$\bar{3}$m
E2$_1$	CaO_3Ti	cP5	Pm3m				
L1$_2$	$AuCu_3$	cP4	Pm3m				
A15	W_3O	cP8	Pm3n				
C46	$AuTe_2$	oP24	Pma2				
B19	β'AuCd	oP4	Pmma				
D0$_a$	βCu_3Ti	oP8	Pmmn				
D0$_d$	$AsMn_3$	oP16	Pmmn				
C3	Cu_2O	cP6	Pn3m				
Ac	αNp	oP8	Pnma				
B16	GeS	oP8	Pnma				
B27	BFe	oP8	Pnma				
B31	MnP	oP8	Pnma				
C23	$PbCl_2$	oP12	Pnma				
D0$_{20}$	Al_3Ni	oP16	Pnma				
D5$_{10}$	Cr_3C_2	oP20	Pnma				
D5$_8$	S_3Sb_2	oP20	Pnma				
D10$_1$	Cr_7C_3	hP80	Pnma				
E9$_e$	$CuFe_2S_3$	oP24	Pnma				
F5$_6$	CuS_2Sb	oP16	Pnma				
C18	FeS_2	oP6	Pnnm				
B13	α-NiS	hR6	R3m				
D8$_{10}$	Al_8Cr_5	hR26	R3m				

[After T. B. Massalski (Editor), *Binary Alloy Phase Diagrams*, **2**, ASMI, Metals Park, 1986, p. 2170; by permission.]

10
Convergent Beam Electron Diffraction

10.1. Introduction

Developments in instrumentation over the past few years have resulted in a surge of activity in studies using convergent beam electron diffraction (CBED). The fundamental studies of CBED techniques and interpretation are far from complete, making it impossible to include complete details in this book. Lengthy descriptions of methods and applications are given in the books by Tanaka and Terauchi [1985] and Tanaka, et al. [1988]. The subjects presented in this chapter cover details not explicitly covered elsewhere.

Convergent beam electron diffraction is produced by converging the beam by exciting the condenser lens strongly in order to produce a nonparallel beam converged to a small diameter. The resulting diffraction pattern has high spatial resolution (small beam diameter), but it has low angular resolution because of the spread of the diffraction spots to discs.

Information obtainable from a CBED pattern includes the following:
- specimen thickness;
- lattice spacing along the zone axis;
- symmetry data related to the point group of the crystal;
- symmetry data related to the space group of the crystal;
- qualitative conclusions concerning stresses present;
- qualitative conclusions concerning changes in lattice parameter;
- qualitative and quantitative conclusions concerning interface orientations.

The formation of discs of intensity as opposed to points is illustrated in Figures 10.1 and 10.2.

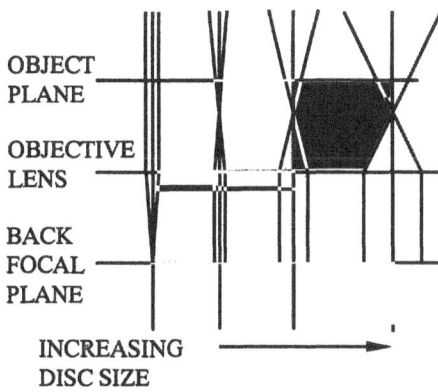

OBJECT PLANE

OBJECTIVE LENS

BACK FOCAL PLANE

INCREASING DISC SIZE

Figure 10.1. As the convergence of the beam increases, the point at the back focal plane of the objective lens is enlarged into a disc, the diameter of which is proportional to the convergence angle.

127

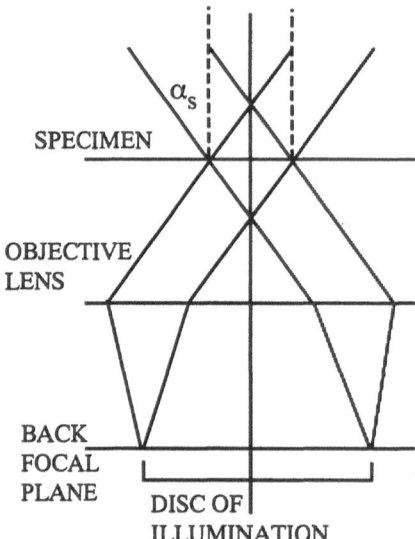

SPECIMEN

OBJECTIVE LENS

BACK FOCAL PLANE

DISC OF ILLUMINATION

Figure 10.2. The truncated cone of the beam forms a disc at the back focal plane of the objective lens.

10.2. Problems with Obtaining and Interpreting the Patterns

10.2.1. Instrument and specimen related problems

The pattern is sensitive to microscope alignment; stage movement is not sensitive enough in tilt; contamination reduces contrast in seconds to minutes; the thickness is too large or too small.

10.2.2. Recording the pattern

Since the intensity range is very wide, the photographic emulsion cannot accommodate the pattern intensities, thereby requiring a series of photos for each pattern; special developing procedures may be required.

10.2.3. Interpretation

If the lattice of the phase is not known, lattice parameters may not be known, resulting in examination of many possibilities; many patterns from a number of zones are required for identification of an unknown phase; the rules for identifying symmetry elements are complex when the space group is desired; point group interpretations are more straightforward, but identifying symmetry present in the pattern and applying the methods takes practice.

CBED requires a small focused probe to provide a small sampling volume and to minimize thickness and tilt variations. The CBED pattern is a two-dimensional map of the diffracted intensity as a function of inclination between the

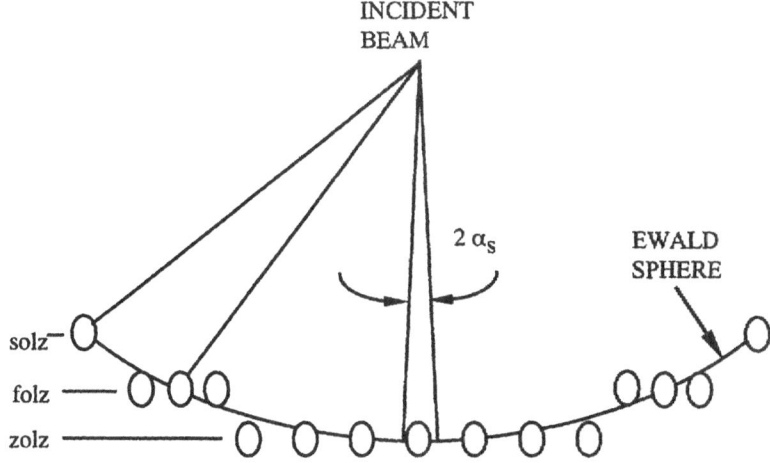

Figure 10.3. The geometry of the reciprocal lattice and the intersection with the Ewald sphere forming higher order Laue zones and the zero order Laue zone.

incident electron beam and a crystal lattice direction. This pattern consists of discs due to Bragg reflection. Intensity variations within the discs contain information about specimen orientation, thickness, symmetry, and structure.

10.3. Zero and Higher Order Laue Zones

Laue zones are defined as those sets of reciprocal lattice points that satisfy the equation

$$\mathbf{g} \cdot \mathbf{u} = N, \tag{10.1}$$

where $\mathbf{g}$ is the reciprocal lattice vector, $\mathbf{u}$ is the zone axis vector, and N is an integer $= 0, \pm1, \pm2, \pm3, \ldots$. The zero order Laue zone, zolz, is the pattern usually obtained in the selected area diffraction mode. It represents the intersection of the Ewald sphere with the origin of reciprocal space and planes in the zone $\mathbf{u}$.

The higher order Laue zones, holz, represent the set of reciprocal lattice points that intersect the Ewald sphere and do not lie in the zolz. The pattern obtained consists of the zolz and a set of rings. Figure 10.3 shows the formation of three Laue zones produced by the intersection of the Ewald sphere with sets of reciprocal lattice points.

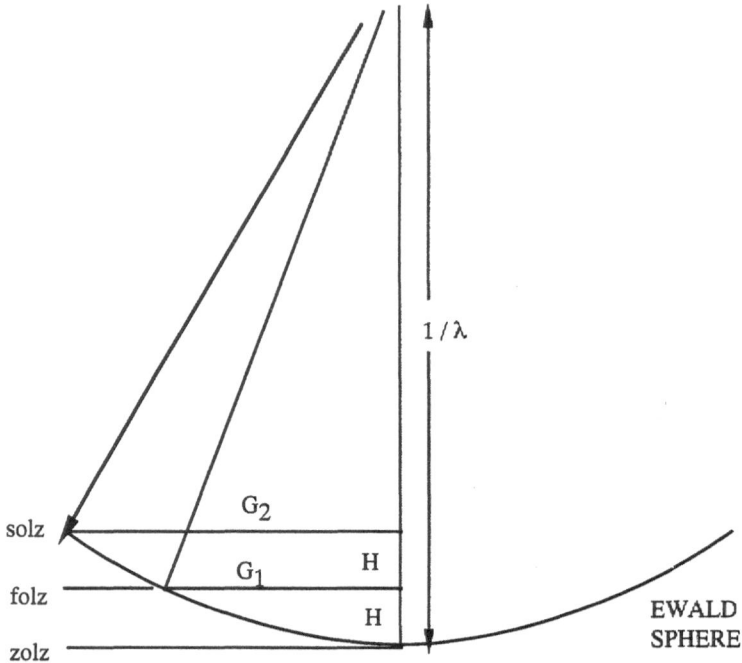

Figure 10.4. The geometry of the higher order Laue zones from which the relationship between G and H is derived.

10.4. Lattice Parameter Along the Zone Axis

The holz contain information on the planar separation in the beam direction, i.e., along the zone axis when oriented appropriately.

The distance, G, from the (000) beam to the ring can be written in terms of the separation of the Laue zone from the zolz and the radius of the Ewald sphere, $1/\lambda$. The geometry is shown in Figure 10.4. From the geometry in Figures 10.4 and 10.5,

$$G^2 + \left(\frac{1}{\lambda} - H^2\right) = \frac{1}{\lambda^2}.$$ (10.2)

Solving for G and dropping second order terms gives

$$G = \sqrt{\frac{2H}{\lambda}}$$ (10.3)

or the first order Laue zone (folz). For the second order Laue zone (solz), the expression for G is

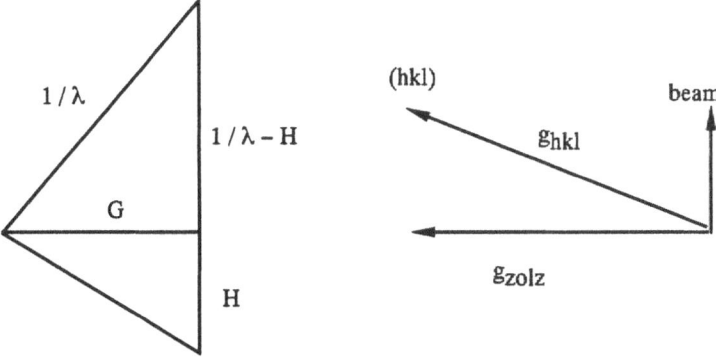

Figure 10.5. Details of the construction used in the derivation of G(H) equation.

$$G = \sqrt{\frac{4H}{\lambda}} = 2\sqrt{\frac{H}{\lambda}},$$ (10.4)

and for the nth zone

$$G = \sqrt{\frac{2nH}{\lambda}}.$$ (10.5)

The value of H, in Å^{-1} or nm^{-1}, is given by the equations in Table 10.8.

10.5. Higher Order Laue Zone Lines in Diffraction Discs

When a convergent beam is present and the camera length is large enough to see detail in the discs, there are lines in the disc that arise from the interaction of the holz with the zolz. This is a dynamic diffraction effect, because although the thickness is large enough for dynamic diffraction to be occurring, it is not so thick that the intensity is completely lost or the electron beam is completely scattered.

Examination of the transmitted disc shows that the lines do not cross at the center of the disc. The displacement of the holz from the zolz produces this failure to converge at the center of the transmitted disc. Let 2ρ = the angle from the (000) to a holz spot, and let θ = the Bragg angle for the holz spot.

Consider the situation in Figure 10.6, which is a diagram showing the geometry of the exact Bragg case where the Ewald sphere and the exact reciprocal lattice point of the plane (hkl) intersect. From the geometry,

$$r = g \cos \theta = \frac{2}{\lambda} \sin \theta \cos \theta = \frac{1}{\lambda} \sin 2\theta,$$ (10.6)

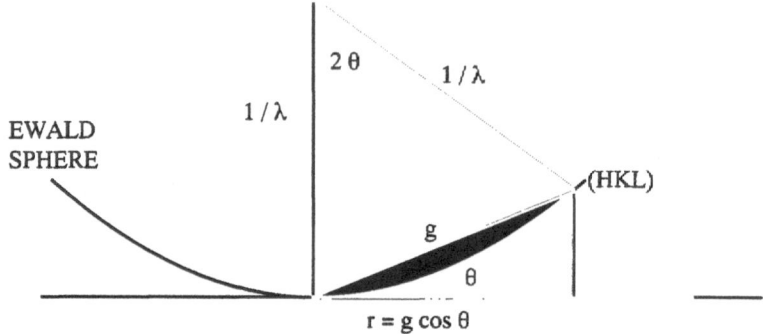

Figure 10.6. Geometry of exact Bragg diffraction condition from holz disc.

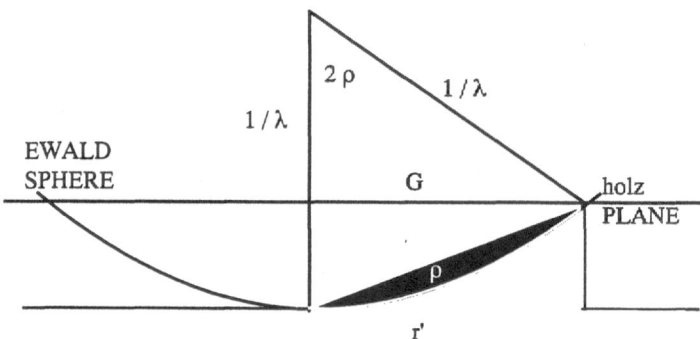

Figure 10.7. Geometry of the radius of the holz ring.

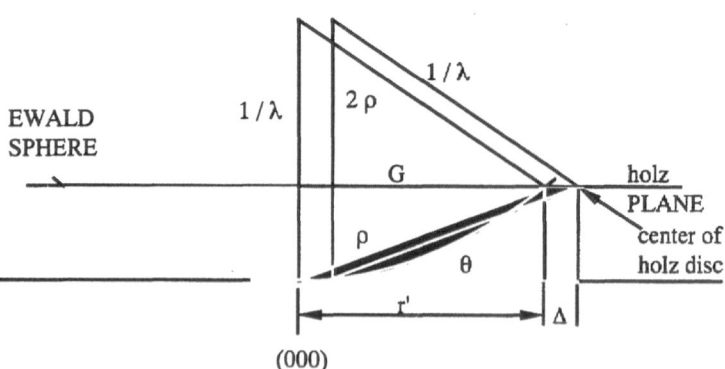

Figure 10.8. Geometry of combined case.

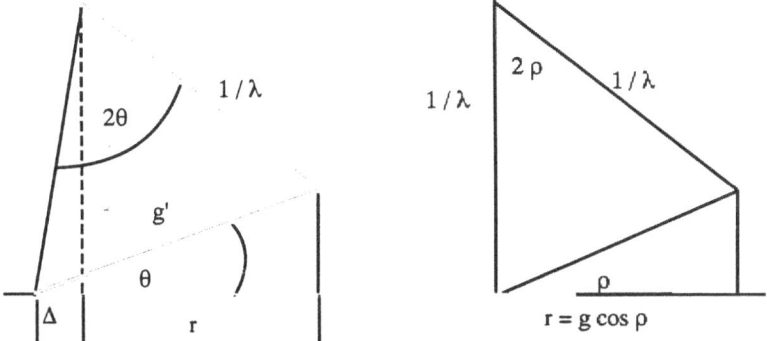

Figure 10.9. left: the displacement from the beam axis. right: the geometry for the line from the center of the transmitted disc to the holz disc center.

where the Bragg equation has been used for the sin θ form. The geometry for the intersection of the Ewald sphere with a holz is shown in Figure 10.7.

The intersection of the Ewald sphere and the holz with a holz reciprocal lattice point or disc is not assumed. From the geometry

$$r' = g' \cos \rho = (2/\lambda) \sin \rho \cos \rho, \tag{10.7}$$

where sin ρ arises from the geometry of the isosceles triangle formed by $1/\lambda$ and g.

In the general case, the holz radius and the exact Bragg center of a holz disc do not coincide, the difference being defined to be Δ, as suggested in Figure 10.8. Now when the reciprocal lattice points are expanded into discs, a portion of the disc may intersect the Ewald sphere, giving rise to an intersection line of intensity and a corresponding deficiency line in the (000) disc. The deficiency line is, in general, displaced from the center of the (000) disc, because the line produced by the intersection of the Ewald sphere and the holz disc is displaced from the center of the holz disc.

To determine the size of the displacement of the line in the (000) disc, the two previous figures must be combined, the one for the exact Bragg case being translated along the holz plane toward the Ewald sphere intersection with the holz plane. This translation can be positive or negative depending on the specific holz disc considered. This combined geometry is shown in Figures 10.9 and 10.10.

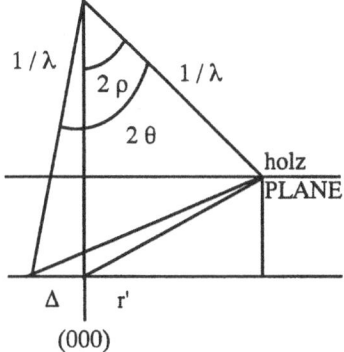

Figure 10.10. Geometry of the displacement of the holz line with respect to the transmitted disc showing the translation made of the point in the holz disc that results in a displacement of the holz line in the zolz disc.

From Figure 10.9,

$$r' - r = \Delta, \tag{10.8}$$

which can be rewritten using the equations above as

$$\Delta = \frac{1}{\lambda} (\sin 2\rho - \sin 2\theta). \tag{10.9}$$

Equation (10.9) can be simplified by noting that the angles are small (<1°) to

$$\Delta = \frac{2}{\lambda} (\rho - \theta). \tag{10.10}$$

If $\theta > \rho$, then $\Delta < 0$, and the holz disc center lies outside the Ewald sphere. If $\rho > \theta$, then $\Delta > 0$, and the holz disc center lies inside the Ewald sphere. We note that these signs correspond to the sign associated with the deviation parameter s.

In summary, then, the displacement of the holz line from the center of the (000) disc is proportional to the difference between the angle from (000) to the holz disc center and the Bragg angle to the holz disc.

Example: <111> FCC zolz, folz

The folz plane 1/d is given by

$$\frac{1}{d} = g = \frac{\sqrt{h^2 + k^2 + l^2}}{a}. \tag{10.11}$$

For specific planes this becomes

plane	1/d (Å)	plane	1/d (Å)
$\bar{7}\bar{1}9$	$\dfrac{11.45}{a}$	$\bar{5}\bar{3}9$	$\dfrac{10.72}{a}$
$\bar{3}\bar{5}9$	$\dfrac{10.72}{a}$	$\bar{1}\bar{7}9$	$\dfrac{11.45}{a}$

For a = 4.00 Å, the radius of the folz ring is

$$G = \sqrt{\frac{2H}{\lambda}} = 2.78 \text{ Å}^{-1}, \tag{10.12}$$

where $\lambda = 0.0372$ Å (100 kV) and

$$H = \frac{1}{a \sqrt{u^2 + v^2 + w^2}} = \frac{1}{a\sqrt{3}} = 0.144 \text{ Å}^{-1}. \tag{10.13}$$

The Bragg angles for (719) and (539) are 0.1066 and 0.09986 rad, respectively. The angle from the holz to the transmitted disc is given by tan (H/G); thus $\rho = 0.05175$ rad.

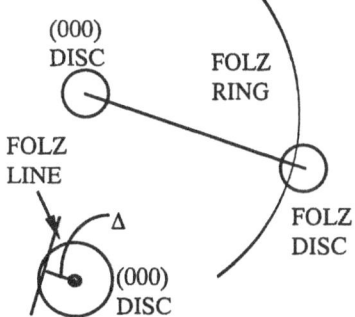

Figure 10.11. Schematic illustration of the displacement of the holz line in the (000) disc.

With this information, we can compare values of g, G, and Δ, and also the sign of the deviation parameter s. Thus

plane	g		G	s	plane	g		G	s
7̄19	2.86	>	2.78	–	3̄5̄9	2.68	<	2.78	+
5̄39	2.68	<	2.78	+	1̄7̄9	2.86	>	2.78	–

From this comparison, it is clear that the the deviation from the center of the transmitted disc has the same sign as the deviation parameter s.

Holz lines are indexed using the folz indexing and applying the proper sign to the value of Δ. As drawn in Figure 10.11, the center of the holz disc is outside the folz ring. Hence, the deviation parameter, s, is negative, because the disc lies outside the Ewald sphere. Therefore, Δ is negative and so lies on the negative side of (000) as measured along the line connecting the holz disc and (000) disc.

10.6. Symmetry Identification

10.6.1. Some definitions

The basic symmetry elements of point groups are the mirror, m, and the five rotational axes, 1, 2, 3, 4, 6. The screw axes and the glide operations constitute the three-dimensional symmetry operations. The zolz exhibits only the symmetry characteristic of the zone axis. Therefore, to identify the crystal symmetry, one must gather data on a large number of zone axes. The holz, however, contain crystal symmetry information, which can be observed in the (000) disc as lines. The diffracted discs may also exhibit symmetries in the intensity distribution within the disc. Hence, although identifying the crystal symmetry requires considerably fewer patterns, it requires careful attention to the pattern in each CBED pattern disc.

Some definitions include the following:

Bright field = the (000) disc.
Whole field or pattern = the entire CBED pattern (zolz + holz).
Dark field = any (hkl) disc.

10.6.2. Point and space group determination

Point group determination is done using observations of the symmetry of the whole field and the bright field. For two simple examples see Stoter [1984].

Kinematically forbidden lines may have an intensity distribution in the form of lines of zero intensity. These lines arise from dynamic effects and are related to the presence of 2_1 screw axes, glide planes, or both. Gjonnes and Moodie [1965] reported on the nature of these lines, now called GM lines. The notation used for GM lines is as follows:

A_2 = the line (in the forbidden disc) produced by interference arising from the mirror reflection of double diffraction paths in the zolz plane, the mirror being perpendicular to the zolz and to the g vector.

B_2 = the line (in the forbidden disc) produced by interference arising from the mirror reflection of double diffraction paths in the zolz, the mirror being perpendicular to the zolz and parallel to the g vector.

The paths are illustrated in Figure 10.12.

Interference can also occur between the zolz and the holz, leading to A_3 (glide effect) and B_3 (screw effect), and intersection of both to give A_3B_3. The A_2B_2 lines are observed when the disc is exactly excited, whereas the A_2 line appears for non-exact conditions. It is present simultaneously in $-g$ as A_2 and in $+g$ as A_2B_2, for $+g$ at exact Bragg condition.

Detailed tables for all 230 space groups have been developed by Tanaka, Sekii and Nagasawa [1983, 1984] and these are used for determining the space group based on observations of the CBED pattern, particularly the presence or absence of glide planes or screw axes.Detailed explanations are given in Tanakaet al. [1983, 1984], Tanaka and Terauchi [1985], and Tanaka et al. [1988]. Note: Corrections to Tables 13 and 17 in Tanaka et al. [1984] were published by Tanaka et al. [1984].

10.7. Thickness Measurement Using Higher Order Laue Zone Lines in a Diffraction Disc

The deviation parameter, s, of a reciprocal lattice vector g in terms of measurable parameters is

$$s = \frac{\lambda}{d(hkl)^2} \frac{\Delta\theta}{2\theta_B},$$ (10.14)

where θ_B is the Bragg angle for g, d is $1/|g|$, and $\Delta\theta$ is the angular separation of lines in the disc. The thickness, extinction distance, and deviation parameter are related through the equation

$$(s^2 + \frac{1}{\xi_g^2})\, t^2 = n^2,$$ (10.15)

where n is an integer. Combining equations we obtain

$$\frac{s_i^2}{n_i^2} = \frac{1}{t^2} - \frac{1}{\xi_g^2\, n_i^2}.$$ (10.16)

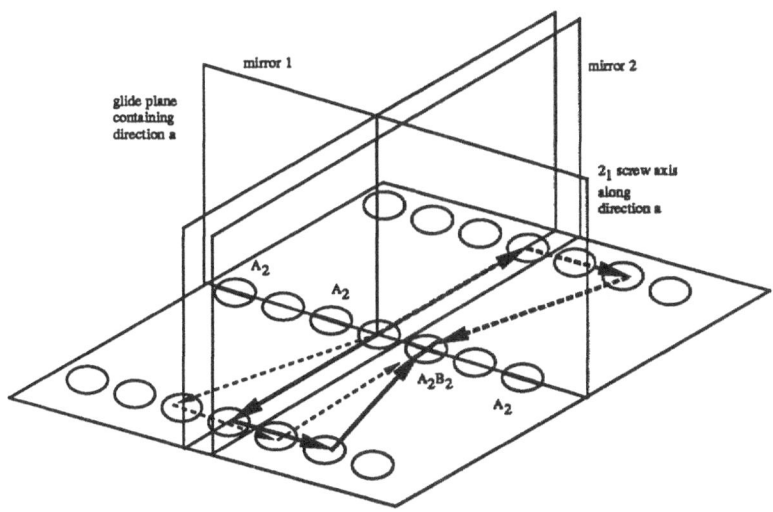

Figure 10.12. The diffraction discs in a CBED pattern are shown with the pattern oriented to exact Bragg condition for the (100) forbidden disc indicated by A_2B_2 and displaying a cross. The GM line A_2 is produced by the scattering along the solid and thick-dotted arrows, which are reflections in the mirror plane m_1 containing direction **a** and a glide plane. A_2 lies parallel to the g vector. The B_2 line is produced by the solid arrow path and the path represented by the fine dotted arrow path, a reflection of the solid path in mirror 2. B_2 lies perpendicular to the **g** vector. Mirror 1 also contains the 2_1 screw axis. See text for further details.

By plotting (s_i^2 / n_i^2) against $1/n_i^2$, the thickness is found by noting the intercept, and the extinction distance is found by calculating the slope of the line. In practice the line may not be linear for the first trial series of n_i. If this occurs, the series is started with $n_i + 1$. This is repeated until a straight line is obtained.

10.8. Indexing Holz Patterns

The method for indexing holz patterns is somewhat more involved than indexing SAD patterns, because the holz pattern lies above the zolz. The procedure relies on the assumption that the zolz has been indexed and the zone found.

In a holz pattern, the indices can be found by using the zone law and vector addition. For the holz patterns, the zone law is

$$\mathbf{g} \cdot \mathbf{u} = N, \tag{10.17}$$

where N is an integer corresponding to the Laue zone number. After the zolz indices are determined, the folz indices can be found by identifying one beam in the folz and then generating the others by vector addition of this beam with each beam in the zolz.

As an example consider the <111> zone in an FCC lattice. The planes in the zolz are listed Table 10.1 together with the sum with the $(11\bar{1})$ plane which lies in the

Table 10.1. Sums of zolz and a folz Vector $(11\bar{1})$ to produce g vectors in the FCC Folz.

g	$+(11\bar{1})$	g	$+(11\bar{1})$	g	$+(11\bar{1})$
$(02\bar{2})$	$(13\bar{3})$	$(\bar{2}20)$	$(\bar{1}3\bar{1})$	$(\bar{2}02)$	$(\bar{1}11)$
$(0\bar{2}2)$	$(1\bar{1}1)$	$(2\bar{2}0)$	$(3\bar{1}\bar{1})$	$(20\bar{2})$	$(31\bar{3})$

folz, because $(11\bar{1}) \cdot (111) = 1$. However, note that if the **g** vectors in the zolz are simply added to the zone axis vector, the resulting vectors do not lie in the folz.

We have done two things: first, we have identified a plane in the folz by using the zone law; second, we have established the identity of planes in the folz by adding zolz planes to the one folz plane. The question is where do the folz planes lie with respect to the zolz planes?

When a crystal structure is viewed along some arbitrary direction, the atoms in each plane perpendicular to that direction may or may not lie exactly above each other. In FCC, for example, the stacking along [111] is ABC, i.e., three equivalent planes, the atoms of which occupy specific positions shifted with respect to the first plane designated as A. Figure 10.13 illustrates the stacking of the adjacent layers. If the x and o atom positions in each plane were shifted by some amount so that all were to lie over the center marked by the square then the stacking would be ...AAA.... . This shift can be calculated by noting the geometry of the structure.

Figure 10.14 is a projection of an arbitrary plane in a structure in the direct lattice. **D** = vector defining the direction **D** perpendicular to the plane P. **d** = vector to the atom position with coordinates [u,v,w]. **t** = vector from the origin to a point on P where the projection of **d** terminates. **R** = vector projection of **d** onto **D**.
θ = the angle between **d** and **R**.

From the geometry shown in Figure 10.14,

$$t = d - \frac{(d \cdot D)}{|D|^2} D. \tag{10.18}$$

It is convenient to write t as a fraction of a vector to an atom position in the reference plane P,

$$t = f\, r_0. \tag{10.19}$$

In terms of indices,

$$t = f\, [u,v,w] r_0. \tag{10.20}$$

Hence, **t** is the projection of an atom position onto P and is written in terms of an atom position in plane P.

Example: Cubic system.
1. For **D** = [110], **d** = [100]:

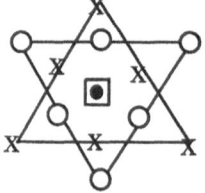

Figure 10.13. Stacking of FCC [111]. x, o represent different levels.

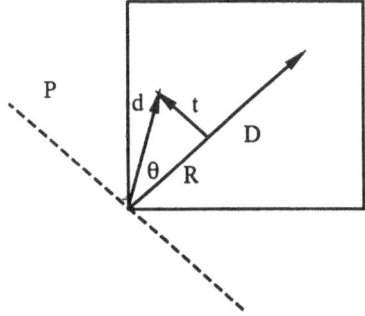

Figure 10.14. Geometry of arbitrary plane and the projection onto a reference plane.

$$t = [100] - \frac{1}{2}[110] = \frac{1}{2}[1\bar{1}0].$$ (10.21)

Hence, f is 1/2 and r_0 is $[1\bar{1}0]$.

2. For $D = [111]$, $d = [1\bar{1}1]$:

$$t = [1\bar{1}1] - \frac{1}{3}[111] = \frac{2}{3}[1\bar{2}1].$$ (10.22)

Hence, f is 2/3, r_0 is $[1\bar{2}1]$. Similar considerations hold for the reciprocal lattice, except that one is dealing with planes rather than atom positions.

The utility of the equations lies in the capability to calculate the projection of atom positions or planes onto a reference plane, from which one can construct a projection of a structure showing atom positions or a projection of planes in reciprocal space onto a reference plane. In reciprocal space, these projections allow one to identify the planes in the holz and where they are located relative to the zolz. To locate beams for any crystal system, one must examine the geometry of the holz patterns, as suggested in Figure 10.15.

From the geometry,

$$t = g_L - u^* H',$$ (10.23)

where g_L = the lattice vector from the origin in the zolz to some $(hkl)_L$ lattice point in zone L; H' = reciprocal lattice separation of Laue zones in units of reciprocal length; u^* = the vector perpendicular to the zolz and parallel to u. It represents a plane.

The unit vector

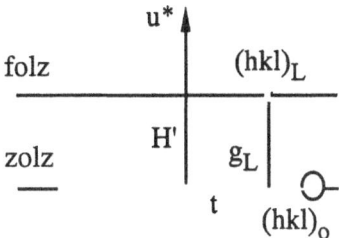

Figure 10.15. Geometry of holz.

$$\hat{u}^* = \frac{u^*}{|u^*|} \qquad (10.24)$$

is used in the following equations.
    For the cubic case

$$\hat{u}^* H' = u^* \frac{H}{|u^*|} = \frac{1}{u^2 + v^2 + w^2} u^*. \qquad (10.25)$$

The equation for **t** can be rewritten as

$$t = g_L - u^* \frac{H}{|u^*|}, \qquad (10.26)$$

which allows calculation of **t** using known indices. Now **t** lies in the zolz along the vector g_0 connecting the origin and a zolz plane $(hkl)_0$. Hence, $|\,t\,|$ can be represented as some fraction of the g_0 and the direction is that of g_0. This is done by writing **t** as

$$t = f\, g_0. \qquad (10.27)$$

In terms of indices

$$t = f\,[(hkl)_L - (u^*v^*w^*)\frac{H\,n}{|u^*|}]. \qquad (10.28)$$

Example for cubic system:
    Given the zone $u = [110]$. For the folz $g \cdot u = 1$. A plane in the folz is (100) since this satisfies the zone law. The **t** vector is

$$t = (100) - \frac{1}{2}(110) = \frac{1}{2}(1\bar{1}0), \qquad (10.29)$$

which means that the projection of the (100) reciprocal lattice point onto the zolz is located along the $(1\bar{1}0)$ **g** vector and is at a distance of $g/2$. With a magnitude and a direction established, the locations of all the planes in the folz are established, because the folz pattern is the same as the zolz except for a general shift of the origin resulting from the structure. The objective of locating the folz pattern relative to the zolz has been accomplished. However, the identity of each disc in the holz pattern remains to be done.

10.9. Construction of the Holz Pattern and Identification of Planes in the Holz and the Holz Ring

The discs in a holz have the same relative positions in the holz as they do in the zolz. The absolute position may be shifted relative to the zolz. The first task, therefore, is to determine whether or not a shift is present. To do this for the folz, use the zone law,

$$hu + kv + lw = 1, \qquad (10.30)$$

to find indices of planes in the folz, g_1. Calculate **t** for one of the folz planes. This will establish the location of g_1 relative to the origin of the zolz.

Superimpose the folz pattern on the zolz pattern and shift the folz pattern as required by **t**.

Example: cubic [100]. Planes in this zone are of {1kl} type. So (100) satisfies the zone law. Thus,

$$t = g_1 - u^* \frac{H}{|u^*|} = (100) - (100) = 0. \tag{10.31}$$

Hence, for this zone there is no shift of the folz.

Example: cubic [110]: Planes in the folz satisfy $h + k = 1$. $g_1 = (011)$ satisfies the law. Thus,

$$t = (011) - \frac{1}{2}(110) = \frac{1}{2}(\bar{1}12). \tag{10.32}$$

Hence, there is a shift of the folz along $(\bar{1}12)$ by $1/2 |(\bar{1}12)|$.

Example: cubic [111]: Planes satisfy $h + k + l = 1$. $g_1 = (1\bar{1}1)$ satisfies the law.

$$t = (1\bar{1}1) - (111)\frac{1}{3} = \frac{2}{3}(1\bar{2}1). \tag{10.33}$$

To index the planes in the folz use vector addition of the one identified g_1 and the labeled vectors of the zolz.

The next problem is to identify the planes that will intercept the Ewald sphere in each holz. From earlier work

$$G = \sqrt{\frac{2H}{\lambda}}. \tag{10.34}$$

Hence, if H can be calculated, then G can be found. But

$$G = \frac{1}{d(hkl)}. \tag{10.35}$$

So one can obtain an equation of a circle of radius G in terms of unknown sets of (hkl) and the known lattice parameters. By calculating d using trios of hkl values, those satisfying the requirement of equality with G can be found.

Example: Cubic system. For cubic systems,

$$H = \frac{1}{a\sqrt{u^2 + v^2 + w^2}}. \tag{10.36}$$

The zone is known since the zolz has been identified and indexed. So H is a known value. Using this information and the relationship between G and H we have, for cubics only,

$$G^2a^2 = h^2 + k^2 + l^2, \tag{10.37}$$

where the (hkl) must satisfy the zone law for the holz under consideration.

For the folz and the [1$\bar{1}$0] zone, 200 kV beam (λ = 0.00256 nm) and a = 0.30 nm, and noting from the zone law and the given zone that h − k = 1, and l is any integer, one obtains the values listed in Table 10.2.

So for cubic [1$\bar{1}$0] zone, the planes closest to G and forming a ring will consist of the planes listed above and their symmetrically derived sets, which satisfy the zone law, e.g., (984) and (98$\bar{4}$) but not ($\bar{9}$8$\bar{4}$) and ($\bar{9}$8$\bar{4}$).

10.10. Rings in Convergent Beam Diffraction

The holz rings have a radius of G [nm^{-1}] given by

$$G = \sqrt{\frac{2H}{\lambda} L}, \tag{10.38}$$

where L = integer, λ = wavelength, and H is the spacing between planes in reciprocal space. A table of ratios of G for various holz shows that there is some ambiguity in using rings alone to find H experimentally, because

$$\frac{G_2}{G_1} = \sqrt{2} = \frac{G_4}{G_2} = \frac{G_6}{G_3} = \cdots \tag{10.39}$$

and

$$\frac{G_3}{G_1} = \sqrt{3} = \frac{G_6}{G_2} = \frac{G_9}{G_3} = \cdots. \tag{10.40}$$

As an example, suppose a pattern from Nb is obtained that has been identified as a [111] zone. The Laue zones are given in general by

$$hu + kv + lw = L, \tag{10.41}$$

for plane (hkl) and zone [uvw]. L is an integer. Since Nb is BCC with a lattice parameter of a= 0.330 nm, H is expected to be given by

$$H = \frac{1}{a\sqrt{u^2 + v^2 + w^2}}, \tag{10.42}$$

which for [111] is

Table 10.2. Values of h, k, and l for which $h^2 + k^2 + l^2$ is near G^2a^2 = 165.73 (for a = 0.3 nm. $H = \dfrac{1}{3\sqrt{2}}$ = 0.2357 nm^{-1}; G^2 = 1.841 nm^{-2}).

$h^2 + k^2 + l^2$	h	k	l	$h^2 + k^2 + l^2$	h	k	l	$h^2 + k^2 + l^2$	h	k	l
161	9	8	4	170	9	8	5	162	8	7	7
177	8	7	8	166	7	6	9	161	6	5	10
162	5	4	11	169	4	3	12	157	3	2	12
149	2	1	12	174	2	1	13	170	1	0	13
181	10	9	0								

$$H = \frac{1}{a\sqrt{3}} = 1.7495 = \frac{1}{0.57158}\ nm^{-1}. \tag{10.43}$$

Therefore, for a 200 kV beam ($\lambda = 0.00256$ nm), G is predicted to be

$$G = 0.3697\sqrt{L}\ nm^{-1}. \tag{10.44}$$

The pattern radius is measured on the plate and is found to be 0.523 nm^{-1}. Hence, to identify the ring, we take the ratio of the unknown G to the known G and find that

$$\frac{G_x}{G_1} = \frac{0.523}{0.3697} = 1.42 = \sqrt{2}. \tag{10.45}$$

From this we conclude that the ring is from the second Laue zone, not the first. This extinction of the first Laue zone for the [111] zone orientation results from the restrictions placed on the BCC indices, i.e., the sum of the indices must be even:

$$h + k + l = 2n,\ [uvw] = [111]; \tag{10.46a}$$
$$h + k + l = 1\ \text{is not allowed}; \tag{10.46b}$$
$$h + k + l = 2,4,6,\ \dots\ \text{is allowed}. \tag{10.46c}$$

What about [-111]? Do these rules hold for it as well? The Laue equation is

$$-h + k + l = 1. \tag{10.47}$$

The possible odd/even combinations of the indices are as follows:

If h = odd, then k + l = 1 - odd = even; but h(odd) + (k + l)(even) odd, which is not allowed.

If h = even, then k + l = 1 - even = odd; but h(even) + (k+ l)(odd)= odd, which is not allowed.

Thus, the possible values for k and l, for h = any integer, are confined to k + l = 1. So k, l must be odd + even or even + odd, but such combinations violate the sum rule that the sum of the indices must be even. Hence the L = 1 Laue zone will not be present.

Any combination of (hkl) regardless of sign must satisfy the sum rule. Therefore, for any <111> zone, the odd Laue zones will not appear. For other zones, similar considerations must be made. For example, given the [100] zone, we have, from the Laue zone equation, h = 1, k = 0, l = 0:

$$h = 1;\ k, l = \text{any integer}. \tag{10.48}$$

Now since h = 1, from the sum rule we have

$$1 + k + l = 2n, \tag{10.49}$$

or

$$k + l = 2n - 1 = \text{odd integer}. \tag{10.50}$$

We set up a table to systematically examine the possibilities:

h	k	l	Sum = Integer	L = 1
odd	odd	even	even	odd
odd	even		odd	even
	odd			

From this parity table we see that the sum of the indices is always even for this [100] zone, and k + l is always odd, which means that the L = 1 Laue zone is permitted.

For L = 2, we have h = 2 for the Laue equation, with k, l any integers, and

$$k + l = 2n - 2 = \text{even integer.} \tag{10.51}$$

So the parity table for the second Laue zone is as follows:

h	k	l	Sum = Integer L = 2
even	odd	odd	even
	even		
even	even	even	even
	even		

Hence, the second Laue zone is permitted. These considerations can be extended to include any L value, the result being that all Laue zones are present for the [100] zone.

A second example is for the [110] zone. The Laue equation is in this case written as

$$h + k + l \cdot 0 = h + k = L, \tag{10.52}$$

and in this case we have h = L − k. L can be odd or even; therefore, we consider first L = odd integer. Then h = odd − k, which means that the h, k parities must be mixed, i.e., when k = odd, h must be even, and when k = even, h must be odd.

The parity table is as follows:

h	k	l	Sum = Integer	Laue Zone
odd	even	odd	even	odd
even	odd	odd	even	odd
odd	even	even	odd	odd
even	odd	even	odd	odd

From the Sum = Integer column, we see that when l = odd and h, k are not the same parity, the Laue zone will appear. If l = even and the h, k are not the same parity the planes will not appear, and they will not contribute to the odd Laue zone.

Suppose we have L = even. Then h = even − k. So when k = odd, h must be odd, and when k = even, h must be even, i.e., the indices must have the same parity; they cannot be mixed. In this case the parity table becomes the following:

h	k	l	Sum = Integer	Laue Zone
even	even	odd	odd	even
odd	odd	odd	odd	even
even	even	even	even	even
odd	odd	even	even	even

From the Sum = Integer column, the only planes to appear are those with all even indices or h, k, odd, l even. In these cases, only even Laue zones appear, as anticipated from our starting assumptions. The other cases for l = odd do not contribute to the Laue zones because the planes do not appear.

A third example is the [120] zone, for which the Laue equation is

$$h + 2k + l \cdot 0 = h + 2k = L. \tag{10.53}$$

The parity table is as follows:

h	k	l	Sum	L
odd	odd	odd	odd	odd
odd	even	odd	even	odd*
odd	odd	even	even	odd*
odd	even	even	odd	odd
even	odd	odd	even	even*
even	even	odd	odd	even
even	odd	even	odd	even
even	even	even	even	even*

The rows with * indicate that the planes are present. Therefore, both even and odd Laue zones appear for the [120] zone.

However, for the [210] zone:

$$2h + k = L. \tag{10.54}$$

The parity table is as follows:

h	k	l	sum	L
odd	odd	odd	odd	odd
odd	even	odd	even	even*
odd	odd	even	even	odd*
odd	even	even	odd	even
even	odd	odd	even	odd*
even	even	odd	odd	even
even	odd	even	odd	odd
even	even	even	even	even*

For the [210] zone, even and odd Laue zones appear, as in the case for the [120] zone, but the index combinations of h and k are not the same.

For a fourth example, consider the [112] zone. The parity table is as follows:

h	k	l	Sum	L
odd	odd	odd	odd	even
odd	odd	even	even	even*
odd	even	odd	even	odd*
odd	even	even	odd	odd
even	odd	odd	even	odd*
even	odd	even	odd	odd
even	even	odd	odd	even
even	even	even	even	even*

The table was obtained by applying the Laue equation

$$h + k + 2l = L. \qquad (10.55)$$

The appearance of Laue rings can be summarized using the results from the parity table as: h, k, l all even, or two even and one odd. Other combinations produce plane indices, which do not appear and so do not contribute to the Laue ring. In summary, this approach of using the parity table to identify which Laue zones will appear yields information on
1. which values of L will be allowed for given parities of indices;
2. which indices will be allowed in a zone of given parity; parity conditions are explicitly identified for each set of indices given the parity of the zone.

10.11. Interpretation of hcp CBED Ring Patterns

In conducting an analysis of convergent beam rings from hexagonal or trigonal lattices, the calculated value of H versus the measured value often differs by a factor of three [Raghavan, Scanlon, and Steeds, 1984].
The root of the problem lies in the fact that the reciprocal lattice derived from the four axis, four index notation fails to fulfill the requirements for a reciprocal lattice as pointed out in Section 5.8. The two expressions for the layer spacing in reciprocal space serve to clarify the source of the difficulty. The equations are

$$H = H_{(4, 4)} = \frac{1}{a\sqrt{3\,(u^2 + v^2 + uv) + \gamma^2\,w^2}}, \qquad (10.56)$$

and

$$H = H_{(3, 3)} = \frac{1}{a\sqrt{u^2 + v^2 - uv + \gamma^2\,w^2}}, \qquad (10.57)$$

where u, v, w and U, V, W are the direction indices, a and c are the lattice parameters, and γ is c/a.
From the conversion equations,

$$U^2 + V^2 + UV = 3(u^2 + v^2 - uv), \qquad (10.58)$$

that is, the magnitude of $|\mathbf{u}|$ is the same in (3, 3) or (4, 4).
Considering the conversion equations

$$u = (1/3)(2U - V), \qquad v = (1/3)(2V - U), \qquad (10.59)$$

where u, v are integers, one obtains a factor of three for some directions. For example, the $[2\overline{1}\overline{1}0]$ direction is parallel to the [100] direction, but the equivalent vectors are $[2\overline{1}\overline{1}0]$ and [300], not [100]. Calculating the H values for $[2\overline{1}\overline{1}0]$ and [100] gives

$$H[2\overline{1}\overline{1}0] = \frac{1}{3a} \text{ and } H[100] = \frac{1}{a}, \qquad (10.60)$$

which differ by a factor of three. Consequently, the H values calculated in the two notations are numerically equal only when the u, v satisfy the mod 3 requirements imposed by the equations relating them to the U, V of the (3, 3) notation. which means that, given a U, V set, if u, v are integers divisible by 3, then the two H values are numerically equal.

To explain the experimental observation $H_{exp} = 3H_{theor}$ for some planes, and $H_{exp} = H_{theor}$ for other planes a detailed list of plane indices in (4, 4) and (3, 3) are presented in Tables 10.3 and 10.4. From this Table one concludes that the presence of a factor of three in the (3, 3) indices leads to an experimental H value equal to three times the theoretical value. When no common factor of three is present the experimental and the theoretical values of H are equal.

Examples of these observations and rules are listed in Table 10.3 for experimentally observed convergent beam ring patterns from TiB_2 [Jackson and Saqib, 1988], which is hexagonal, has space group P6/mmm, and has no general or specific extinctions. The ±5% agreement between the measured and the calculated values is reasonable, while the agreement with the rules is exact.

For reference purposes when analyzing convergent beam electron diffraction patterns, Tables 10.5 through 10.12 have been collected at the end of this Chapter, because this is the most useful and convenient location for the information they contain.

Table 10.3. Experimentally Observed and Calculated Values of H for Various hcp Planes of TiB_2.

(4, 4)	(3, 3)	H_{theor}	H_{exp}
$[2\bar{1}\bar{1}0]$	[300]	x3 = 0.3300	0.3324
$[2\bar{1}\bar{1}3]$	[303]	x3 = 0.2257	0.2205
$[7\bar{5}\bar{2}3]$	$[9\bar{3}3]$	x3 = 0.0878	0.0865
$[5\bar{4}\bar{1}3]$	$[6\bar{3}3]$	x3 = 0.1157	0.1152
$[1\bar{1}01]$	$[1\bar{1}1]$	x1 = 0.1622	0.1637*
$[4\bar{5}16]$	$[3\bar{6}6]$	x3 = 0.0971	0.0972*
$[0\bar{1}12]$	$[\bar{1}22]$	x1 = 0.1201	0.1246*

*Corrected for 4% distortion in pattern.

Table 10.4. Expected Values of H_{exp} in Terms of H_{theor} for Various hcp Planes in (4, 4) and (3, 3) Notation.

(4, 4)	(3, 3)	n	3?	(4, 4)	(3, 3)	n	3?
[0001]	[001]	1	n	$[01\bar{1}\bar{1}]$	$[12\bar{1}]$	1	n
$[11\bar{2}0]$	[330]	3	y	$[\bar{1}101]$	$[\bar{1}11]$	1	n
$[\bar{1}2\bar{1}0]$	[030]	3	y	$[20\bar{2}3]$	[423]	1	n
$[2\bar{1}\bar{1}0]$	[300]	3	y	$[02\bar{2}3]$	[243]	1	n
$[\bar{1}2\bar{1}1]$	[031]	1	n	$[\bar{2}203]$	$[\bar{2}23]$	1	n
$[\bar{1}2\bar{1}6]$	[036]	3	y	$[10\bar{1}2]$	[212]	1	n
$[11\bar{2}3]$	[333]	3	y	$[01\bar{1}2]$	[122]	1	n
$[\bar{1}2\bar{1}3]$	[033]	3	y	$[1\bar{1}02]$	$[1\bar{1}2]$	1	n
$[\bar{2}113]$	[303]	3	y	$[12\bar{3}0]$	[450]	1	n
$[\bar{2}4\bar{2}3]$	[063]	3	y	$[5\bar{1}\bar{4}0]$	[930]	3	y
$[10\bar{1}0]$	[210]	1	n	$[5\bar{4}\bar{1}0]$	$[6\bar{3}0]$	3	y
$[01\bar{1}0]$	[120]	1	n	$[5\bar{1}\bar{4}3]$	[933]	3	y
$[\bar{1}100]$	$[\bar{1}10]$	1	n	$[5\bar{4}\bar{1}3]$	$[6\bar{3}3]$	3	y
$[10\bar{1}1]$	[211] • 1		n				

Table 10.5. Forms for H for the Seven Crystal Systems.

Cubic: $\qquad H = \dfrac{1}{a\sqrt{u^2 + v^2 + w^2}}$,

FCC: $\qquad H = \dfrac{p}{a\sqrt{u^2 + v^2 + w^2}}$,

p = 1 if u + v + w = odd; p = 2 if u + v + w = even.

BCC: $\qquad H = \dfrac{p}{a\sqrt{u^2 + v^2 + w^2}}$,

p = 2 if u, v, w all odd; p = 1 if u, v, w are even or mixed.

Orthorhombic: $\quad H = \dfrac{1}{\sqrt{a^2 u^2 + b^2 v^2 + c^2 w^2}}$.

Tetragonal: $\quad H = \dfrac{1}{\sqrt{a^2 (u^2 + v^2) + c^2 w^2}}$.

Hexagonal or Trigonal in (3, 3) notation:

$$H = \dfrac{1}{\sqrt{a^2 (u^2 + v^2 - uv) + c^2 w^2}} .$$

Monoclinic (b axis unique):

$$H = \dfrac{1}{\sqrt{a^2 u^2 + b^2 v^2 + c^2 w^2 + 2\,a\,c\,u\,w\,\cos\beta}} .$$

General or Triclinic: $H = \dfrac{1}{\sqrt{a^2 u^2 + b^2 v^2 + c^2 w^2 + p + q + r}}$.

where $p = 2\,a\,c\,u\,v\,\cos\gamma,\quad q = 2\,b\,c\,v\,w\,\cos\alpha,\quad r = 2\,a\,c\,u\,w\,\cos\beta.$

a, b, c are the lattice parameters; α, β, γ are the axis angles; u, v, w are the zone axis indices.

Table 10.6. Diffraction Groups, Zones, Point Groups, and Crystal Systems Arranged by Diffraction Group.

D Grp	Zone	Pt Grp	CS	D Grp	Zone	Pt Grp	CS
1	<uvtw>	$\bar{6}$	h	$2m_Rm_R$	<1$\bar{1}$00>	622	h
1	<uvtw>	$\bar{6}$m2	h	$2m_Rm_R$	<11$\bar{2}$0>	622	h
1	<uvtw>	3	r	$2m_Rm_R$	<100>	$\bar{4}$2m	t
1	<uvtw>	3m	r	$2m_Rm_R$	<100>	23	c
1	<uvtw>	6	h	$2m_Rm_R$	<100>	222	o
1	<uvtw>	6mm	h	$2m_Rm_R$	<100>	422	t
1	<uvtw>	32	r	$2m_Rm_R$	<110>	422	t
1	<uvtw>	622	h	$2m_Rm_R$	<110>	432	c
1	<uvw>	$\bar{4}$	t	$2m_Rm_R$	[001]	222	o
1	<uvw>	$\bar{4}$2m	t	2_R	[uvtw]	$\bar{3}$	r
1	<uvw>	$\bar{4}$3m	c	2_R	[uvtw]	$\bar{3}$m	r
1	<uvw>	1	a	2_R	[uvtw]	6/m	h
1	<uvw>	2	m	2_R	[uvtw]	6/mmm	h
1	<uvw>	4	t	2_R	[uvw]	$\bar{1}$	a
1	<uvw>	4mm	t	2_R	[uvw]	2/m	m
1	<uvw>	23	c	2_R	[uvw]	4/m	t
1	<uvw>	222	o	2_R	[uvw]	4/mmm	t
1	<uvw>	422	t	2_R	[uvw]	m3	c
1	<uvw>	432	c	2_R	[uvw]	m3m	c
1	<uvw>	m	m	2_R	[uvw]	mmm	o
1	<uvw>	mm2	o	2_Rmm_R	<uuw>	m3m	c
1_R	<11$\bar{2}$0>	3m	h	2_Rmm_R	<uv0>	m3	c
1_R	[010]	m	m	2_Rmm_R	<uv0>	m3m	c
2	<11$\bar{2}$0>	32	r	2_Rmm_R	[0001]	6/m	h
2	[010]	2	m	2_Rmm_R	[u$\bar{u}$0w]	$\bar{3}$m	h
2mm	<1$\bar{1}$00>	$\bar{6}$m2	h	2_Rmm_R	[u$\bar{u}$0w]	6/mmm	h
2mm	[001]	mm2	o	2_Rmm_R	[u0w]	2/m	m
$2mm1_R$	<1$\bar{1}$00>	6/mmm	h	2_Rmm_R	[u0w]	4/mmm	t
$2mm1_R$	<11$\bar{2}$0>	6/mmm	h	2_Rmm_R	[u0w]	mmm	o
$2mm1_R$	<100>	4/mmm	t	2_Rmm_R	[uv0]	4/m	t
$2mm1_R$	<100>	m3	c	2_Rmm_R	[uv0]	4/mmm	t
$2mm1_R$	<100>	mmm	o	2_Rmm_R	[uv0]	mmm	o
$2mm1_R$	<110>	4/mmm	t	2_Rmm_R	[uvt0]	6/mmm	h
$2mm1_R$	<110>	m3m	c	2_Rmm_R	[uvtw]	6/mmm	h
$2mm1_R$	[001]	mmm	o	2_Rmm_R	[uvw]	4/mmm	t

Table 10.6. Continued.

D Grp	Zone	Pt Grp	CS	D Grp	Zone	Pt Grp	CS
3	<111>	$\bar{2}$3	c	m	[uu$\bar{2}$uw]	6mm	h
3	[0001]	3	r	m	[uuw]	$\bar{4}$2m	t
3m	<111>	$\bar{4}$3m	c	m	[uuw]	4mm	t
3m	[0001]	3m	r	m	[uvt0]	$\bar{6}$	h
3m1$_R$	[0001]	$\bar{6}$m2	h	m	[uvt0]	$\bar{6}$m2	h
3m$_R$	<111>	32	r	m1$_R$	<1$\bar{1}$00>	6mm	h
3m$_R$	<111>	432	c	m1$_R$	<11$\bar{2}$0>	$\bar{6}$m2	h
4	[001]	4	t	m1$_R$	<11$\bar{2}$0>	6mm	h
4mm	[001]	4mm	t	m1$_R$	<100>	4mm	t
4mm1$_R$	<100>	m3m	c	m1$_R$	<100>	mm2	t
4mm1$_R$	[001]	$\bar{4}$2m	t	m1$_R$	<110>	$\bar{4}$2m	t
4m$_R$m$_R$	<100>	432	c	m1$_R$	<110>	$\bar{4}$3m	c
4m$_R$m$_R$	[001]	422	t	m1$_R$	<110>	4mm	t
4$_R$	[001]	$\bar{4}$	t	m$_R$	<uuw>	432	c
4$_R$mm$_R$	<100>	$\bar{4}$3m	c	m$_R$	<uv0>	$\bar{4}$3m	c
4$_R$mm$_R$	[001]	$\bar{4}$2m	t	m$_R$	<uv0>	$\bar{2}$3	c
6	[0001]	6	h	m$_R$	<uv0>	432	c
6mm	[0001]	6mm	h	m$_R$	[u$\bar{u}$0w]	32	r
6mm1$_R$	[0001]	6/mmm	h	m$_R$	[u$^-$u0w]	622	h
6m$_R$m$_R$	[0001]	622	h	m$_R$	[u0w]	$\bar{4}$2m	t
6$_R$	<111>	m3	c	m$_R$	[u0w]	2	m
6$_R$	[0001]	$\bar{3}$	r	m$_R$	[u0w]	222	o
6$_R$mm$_R$	<111>	m3m	c	m$_R$	[u0w]	422	t
6$_R$mm$_R$	[0001]	$\bar{3}$m	r	m$_R$	[uu$\bar{2}$uw]	$\bar{6}$m2	h
21$_R$	<11$\bar{2}$0>	$\bar{3}$m	h	m$_R$	[uu$\bar{2}$uw]	622	h
21$_R$	[010]	2/m	m	m$_R$	[uuw]	422	t
31$_R$	[0001]	$\bar{6}$	h	m$_R$	[uv0]	$\bar{4}$	t
41$_R$	[001]	4/m	t	m$_R$	[uv0]	$\bar{4}$2m	t
61$_R$	[0001]	6/m	h	m$_R$	[uv0]	4	t
m	<uuw>	$\bar{4}$3m	c	m$_R$	[uv0]	4mm	t
m	[u$\bar{u}$0w]	$\bar{6}$m2	h	m$_R$	[uv0]	222	o
m	[u$\bar{u}$0w]	3m	r	m$_R$	[uv0]	422	t
m	[u$\bar{u}$0w]	6mm	h	m$_R$	[uv0]	mm2	o
m	[u0w]	4mm	t	m$_R$	[uvt0]	6	h
m	[u0w]	m	m	m$_R$	[uvt0]	6mm	h
m	[u0w]	mm2	o	m$_R$	[uvt0]	622	h

Table 10.7. Diffraction Groups, Zones, Point Groups, and Crystal Systems, Arranged by Point Group.

D Grp	Zone	Pt Grp	CS	D Grp	Zone	Pt Grp	CS
2_R	[uvw]	$\bar{1}$	a	2	[010]	2	m
6_R	[0001]	$\bar{3}$	r	m_R	[u0w]	2	m
2_R	[uvtw]	$\bar{3}$	r	21_R	[010]	2/m	m
21_R	<11$\bar{2}$0>	$\bar{3}$m	h	2_Rmm_R	[u0w]	2/m	m
2_Rmm_R	[u$\bar{u}$0w]	$\bar{3}$m	h	2_R	[uvw]	2/m	m
6_Rmm_R	[0001]	$\bar{3}$m	r	1	<uvtw>	3	r
2_R	[uvtw]	$\bar{3}$m	r	3	[0001]	3	r
1	<uvw>	$\bar{4}$	t	1_R	<11$\bar{2}$0>	3m	h
4_R	[001]	$\bar{4}$	t	1	<uvtw>	3m	r
m_R	[uv0]	$\bar{4}$	t	3m	[0001]	3m	r
1	<uvtw>	$\bar{6}$	h	m	[u$\bar{u}$0w]	3m	r
31_R	[0001]	$\bar{6}$	h	1	<uvw>	4	t
m	[uvt0]	$\bar{6}$	h	4	[001]	4	t
2mm	<1$\bar{1}$00>	$\bar{6}$m2	h	m_R	[uv0]	4	t
$m1_R$	<11$\bar{2}$0>	$\bar{6}$m2	h	41_R	[001]	4/m	t
1	<uvtw>	$\bar{6}$m2	h	2_Rmm_R	[uv0]	4/m	t
$3m1_R$	[0001]	$\bar{6}$m2	h	2_R	[uvw]	4/m	t
m	[u$\bar{u}$0w]	$\bar{6}$m2	h	$2mm1_R$	<100>	4/mmm	t
m_R	[uu$\bar{2}$uw]	$\bar{6}$m2	h	$2mm1_R$	<110>	4/mmm	t
m	[uvt0]	$\bar{6}$m2	h	2_Rmm_R	[u0w]	4/mmm	t
$2m_Rm_R$	<100>	$\bar{4}$2m	t	2_Rmm_R	[uv0]	4/mmm	t
$m1_R$	<110>	$\bar{4}$2m	t	2_R	[uvw]	4/mmm	t
1	<uvw>	$\bar{4}$2m	t	2_Rmm_R	[uvw]	4/mmm	t
$4mm1_R$	[001]	$\bar{4}$2m	t	$m1_R$	<100>	4mm	t
4_Rmm_R	[001]	$\bar{4}$2m	t	$m1_R$	<110>	4mm	t
m_R	[u0w]	$\bar{4}$2m	t	1	<uvw>	4mm	t
m	[uuw]	$\bar{4}$2m	t	4mm	[001]	4mm	t
m_R	[uv0]	$\bar{4}$2m	t	m	[u0w]	4mm	t
4_Rmm_R	<100>	$\bar{4}$3m	c	m	[uuw]	4mm	t
$m1_R$	<110>	$\bar{4}$3m	c	m_R	[uv0]	4mm	t
3m	<111>	$\bar{4}$3m	c	1	<uvtw>	6	h
m	<uuw>	$\bar{4}$3m	c	6	[0001]	6	h
m_R	<uv0>	$\bar{4}$3m	c	m_R	[uvt0]	6	h
1	<uvw>	$\bar{4}$3m	c	2_Rmm_R	[0001]	6/m	h
1	<uvw>	1	a	2_R	[uvtw]	6/m	h
1	<uvw>	2	m	61_R	[0001]	6/m	h

Tables   153

Table 10.7. Continued.

D Grp	Zone	Pt Grp	CS	D Grp	Zone	Pt Grp	CS
$2mm1_R$	<1$\bar{1}$00>	6/mmm	h	$3m_R$	<111>	432	c
$2mm1_R$	<11$\bar{2}$0>	6/mmm	h	m_R	<uuw>	432	c
$6mm1_R$	[0001]	6/mmm	h	m_R	<uv0>	432	c
2_Rmm_R	[u$\bar{u}$0w]	6/mmm	h	1	<uvw>	432	c
2_Rmm_R	[uvt0]	6/mmm	h	$2m_Rm_R$	<1$\bar{1}$00>	622	h
2_R	[uvtw]	6/mmm	h	$2m_Rm_R$	<11$\bar{2}$0>	622	h
2_Rmm_R	[uvtw]	6/mmm	h	1	<uvtw>	622	h
$m1_R$	<1$\bar{1}$00>	6mm	h	$6m_Rm_R$	[0001]	622	h
$m1_R$	<11$\bar{2}$0>	6mm	h	m_R	[u$\bar{u}$0w]	622	h
1	<uvtw>	6mm	h	m_R	[uu$\bar{2}$uw]	622	h
6mm	[0001]	6mm	h	m_R	[uvt0]	622	h
m	[u$\bar{u}$0w]	6mm	h	1	<uvw>	m	m
m	[uu$\bar{2}$uw]	6mm	h	1_R	[010]	m	m
m_R	[uvt0]	6mm	h	m	[u0w]	m	m
$2m_Rm_R$	<100>	23	c	$2mm1_R$	<100>	m3	c
3	<111>	23	c	6_R	<111>	m3	c
m_R	<uv0>	23	c	2_Rmm_R	<uv0>	m3	c
1	<uvw>	23	c	2_R	[uvw]	m3	c
2	<11$\bar{2}$0>	32	r	$4mm1_R$	<100>	m3m	c
$3m_R$	<111>	32	r	$2mm1_R$	<110>	m3m	c
1	<uvtw>	32	r	6_Rmm_R	<111>	m3m	c
m_R	[u$\bar{u}$0w]	32	r	2_Rmm_R	<uuw>	m3m	c
$2m_Rm_R$	<100>	222	o	2_Rmm_R	<uv0>	m3m	c
1	<uvw>	222	o	2_R	[uvw]	m3m	c
$2m_Rm_R$	[001]	222	o	1	<uvw>	mm2	o
m_R	[u0w]	222	o	2mm	[001]	mm2	o
m_R	[uv0]	222	o	m	[u0w]	mm2	o
$2m_Rm_R$	<100>	422	t	m_R	[uv0]	mm2	o
$2m_Rm_R$	<110>	422	t	$m1_R$	<100>	mm2	t
1	<uvw>	422	t	$2mm1_R$	<100>	mmm	o
$4m_Rm_R$	[001]	422	t	$2mm1_R$	[001]	mmm	o
m_R	[u0w]	422	t	2_Rmm_R	[u0w]	mmm	o
m_R	[uuw]	422	t	2_Rmm_R	[uv0]	mmm	o
m_R	[uv0]	422	t	2_R	[uvw]	mmm	o
$4m_Rm_R$	<100>	432	c				
$2m_Rm_R$	<110>	432	c				

Table 10.8. Diffraction Groups, Zones, Point Groups, and Crystal Systems, Arranged by Crystal System.

D Grp	Zone	Pt Grp	CS	D Grp	Zone	Pt Grp	CS
1	$\langle uvw \rangle$	1	a	$2mm1_R$	$\langle 11\bar{2}0 \rangle$	6/mmm	h
2_R	$[uvw]$	$\bar{1}$	a	$m1_R$	$\langle 11\bar{2}0 \rangle$	6mm	h
$2m_Rm_R$	$\langle 100 \rangle$	23	c	1	$\langle uvtw \rangle$	$\bar{6}$	h
$4m_Rm_R$	$\langle 100 \rangle$	432	c	1	$\langle uvtw \rangle$	6	h
4_Rmm_R	$\langle 100 \rangle$	$\bar{4}3m$	c	1	$\langle uvtw \rangle$	622	h
$2mm1_R$	$\langle 100 \rangle$	m3	c	1	$\langle uvtw \rangle$	$\bar{6}m2$	h
$4mm1_R$	$\langle 100 \rangle$	m3m	c	1	$\langle uvtw \rangle$	6mm	h
$2m_Rm_R$	$\langle 110 \rangle$	432	c	31_R	$[0001]$	$\bar{6}$	h
$m1_R$	$\langle 110 \rangle$	$\bar{4}3m$	c	6	$[0001]$	6	h
$2mm1_R$	$\langle 110 \rangle$	m3m	c	$6m_Rm_R$	$[0001]$	622	h
3	$\langle 111 \rangle$	23	c	$3m1_R$	$[0001]$	$\bar{6}m2$	h
$3m_R$	$\langle 111 \rangle$	432	c	2_Rmm_R	$[0001]$	6/m	h
3m	$\langle 111 \rangle$	$\bar{4}3m$	c	$6mm1_R$	$[0001]$	6/mmm	h
6_R	$\langle 111 \rangle$	m3	c	61_R	$[0001]$	61m	h
6_Rmm_R	$\langle 111 \rangle$	m3m	c	6mm	$[0001]$	6mm	h
m_R	$\langle uuw \rangle$	432	c	m_R	$[u\bar{u}0w]$	622	h
m	$\langle uuw \rangle$	$\bar{4}3m$	c	2_Rmm_R	$[u\bar{u}0w]$	3m	h
2_Rmm_R	$\langle uuw \rangle$	m3m	c	m	$[u\bar{u}0w]$	$\bar{6}m2$	h
m_R	$\langle uv0 \rangle$	23	c	2_Rmm_R	$[u\bar{u}0w]$	6/mmm	h
m_R	$\langle uv0 \rangle$	432	c	m	$[u\bar{u}0w]$	6mm	h
m_R	$\langle uv0 \rangle$	$\bar{4}3m$	c	m_R	$[uu\bar{2}uw]$	622	h
2_Rmm_R	$\langle uv0 \rangle$	m3	c	m_R	$[uu\bar{2}uw]$	$\bar{6}m2$	h
2_Rmm_R	$\langle uv0 \rangle$	m3m	c	m	$[uu\bar{2}uw]$	6mm	h
1	$\langle uvw \rangle$	23	c	m	$[uvt0]$	$\bar{6}$	h
1	$\langle uvw \rangle$	432	c	m_R	$[uvt0]$	6	h
1	$\langle uvw \rangle$	$\bar{4}3m$	c	m_R	$[uvt0]$	622	h
2_R	$[uvw]$	m3	c	m	$[uvt0]$	$\bar{6}m2$	h
2_R	$[uvw]$	m3m	c	2_Rmm_R	$[uvt0]$	6/mmm	h
$2m_Rm_R$	$\langle 1\bar{1}00 \rangle$	622	h	m_R	$[uvt0]$	6mm	h
2mm	$\langle 1\bar{1}00 \rangle$	$\bar{6}m2$	h	2_R	$[uvtw]$	6/m	h
$2mm1_R$	$\langle 1\bar{1}00 \rangle$	6/mmm	h	2_R	$[uvtw]$	6/mmm	h
$m1_R$	$\langle 1\bar{1}00 \rangle$	6mm	h	2_Rmm_R	$[uvtw]$	6/mmm	h
$2m_Rm_R$	$\langle 11\bar{2}0 \rangle$	622	h	1	$\langle uvw \rangle$	2	m
21_R	$\langle 11\bar{2}0 \rangle$	3m	h	1	$\langle uvw \rangle$	m	m
$m1_R$	$\langle 11\bar{2}0 \rangle$	$\bar{6}m2$	h	2	$[010]$	2	m
1_R	$\langle 11\bar{2}0 \rangle$	3m	h	21_R	$[010]$	2/m	m

Table 10.8. Continued.

D Grp	Zone	Pt Grp	CS	D Grp	Zone	Pt Grp	CS
1_R	[010]	m	m	$m1_R$	<100>	mm2	t
m_R	[u0w]	2	m	$2m_Rm_R$	<110>	422	t
2_Rmm_R	[u0w]	2/m	m	$m1_R$	<110>	$\bar{4}$2m	t
m	[u0w]	m	m	$2mm1_R$	<110>	4/mmm	t
2_R	[uvw]	2/m	m	$m1_R$	<110>	4mm	t
$2m_Rm_R$	<100>	222	o	1	<uvw>	$\bar{4}$	t
$2mm1_R$	<100>	mmm	o	1	<uvw>	4	t
1	<uvw>	222	o	1	<uvw>	422	t
1	<uvw>	mm2	o	1	<uvw>	$\bar{4}$2m	t
$2m_Rm_R$	[001]	222	o	1	<uvw>	4mm	t
2mm	[001]	mm2	o	4_R	[001]	$\bar{4}$	t
$2mm1_R$	[001]	mmm	o	4	[001]	4	t
m_R	[u0w]	222	o	$4m_Rm_R$	[001]	422	t
m	[u0w]	mm2	o	$4mm1_R$	[001]	$\bar{4}$2m	t
2_Rmm_R	[u0w]	mmm	o	4_Rmm_R	[001]	$\bar{4}$2m	t
m_R	[uv0]	222	o	41_R	[001]	4/m	t
m_R	[uv0]	mm2	o	4mm	[001]	4mm	t
2_Rmm_R	[uv0]	mmm	o	m_R	[u0w]	422	t
2_R	[uvw]	mmm	o	m_R	[u0w]	$\bar{4}$2m	t
2	<11$\bar{2}$0>	32	r	2_Rmm_R	[u0w]	4/mmm	t
$3m_R$	<111>	32	r	m	[u0w]	4mm	t
1	<uvtw>	3	r	m_R	[uuw]	422	t
1	<uvtw>	32	r	m	[uuw]	$\bar{4}$2m	t
1	<uvtw>	3m	r	m	[uuw]	4mm	t
6_R	[0001]	$\bar{3}$	r	m_R	[uv0]	$\bar{4}$	t
3	[0001]	3	r	m_R	[uv0]	4	t
6_Rmm_R	[0001]	$\bar{3}$m	r	m_R	[uv0]	422	t
3m	[0001]	3m	r	m_R	[uv0]	$\bar{4}$2m	t
m_R	[u$\bar{u}$0w]	32	r	2_Rmm_R	[uv0]	4/m	t
m	[u$\bar{u}$0w]	3m	r	2_Rmm_R	[uv0]	4/mmm	t
2_R	[uvtw]	$\bar{3}$	r	m_R	[uv0]	4mm	t
2_R	[uvtw]	$\bar{3}$m	r	2_R	[uvw]	4/m	t
$2m_Rm_R$	<100>	422	t	2_R	[uvw]	4/mmm	t
$2m_Rm_R$	<100>	$\bar{4}$2m	t	2_Rmm_R	[uvw]	4/mmm	t
$2mm1_R$	<100>	4/mmm	t				
$m1_R$	<100>	4mm	t				

Table 10.9. Diffraction Groups, Zones, Point Groups, and Crystal Systems, Arranged by Zone.

D Grp	Zone	Pt Grp	CS	D Grp	Zone	Pt Grp	CS
2mm	$\langle 1\bar{1}00\rangle$	$\bar{6}$m2	h	m	$\langle$uuw$\rangle$	$\bar{4}$3m	c
2mm1$_R$	$\langle 1\bar{1}00\rangle$	6/mmm	h	m$_R$	$\langle$uuw$\rangle$	432	c
m1$_R$	$\langle 1\bar{1}00\rangle$	6mm	h	2$_R$mm$_R$	$\langle$uuw$\rangle$	m3m	c
2m$_R$m$_R$	$\langle 1\bar{1}00\rangle$	622	h	m$_R$	$\langle$uv0$\rangle$	$\bar{4}$3m	c
21$_R$	$\langle 11\bar{2}0\rangle$	$\bar{3}$m	h	m$_R$	$\langle$uv0$\rangle$	23	c
m1$_R$	$\langle 11\bar{2}0\rangle$	$\bar{6}$m2	h	m$_R$	$\langle$uv0$\rangle$	432	c
1$_R$	$\langle 11\bar{2}0\rangle$	3m	h	2$_R$mm$_R$	$\langle$uv0$\rangle$	m3	c
2mm1$_R$	$\langle 11\bar{2}0\rangle$	6/mmm	h	2$_R$mm$_R$	$\langle$uv0$\rangle$	m3m	c
m1$_R$	$\langle 11\bar{2}0\rangle$	6mm	h	1	$\langle$uvtw$\rangle$	$\bar{6}$	h
2	$\langle 11\bar{2}0\rangle$	32	r	1	$\langle$uvtw$\rangle$	$\bar{6}$m2	h
2m$_R$m$_R$	$\langle 11\bar{2}0\rangle$	622	h	1	$\langle$uvtw$\rangle$	3	r
2m$_R$m$_R$	$\langle 100\rangle$	$\bar{4}$2m	t	1	$\langle$uvtw$\rangle$	3m	r
4$_R$m$_R$	$\langle 100\rangle$	$\bar{4}$3m	c	1	$\langle$uvtw$\rangle$	6	h
2mm1$_R$	$\langle 100\rangle$	4/mmm	t	1	$\langle$uvtw$\rangle$	6mm	h
m1$_R$	$\langle 100\rangle$	4mm	t	1	$\langle$uvtw$\rangle$	32	r
2m$_R$m$_R$	$\langle 100\rangle$	23	c	1	$\langle$uvtw$\rangle$	622	h
2m$_R$m$_R$	$\langle 100\rangle$	222	o	1	$\langle$uvw$\rangle$	$\bar{4}$	t
2m$_R$m$_R$	$\langle 100\rangle$	422	t	1	$\langle$uvw$\rangle$	$\bar{4}$2m	t
4m$_R$m$_R$	$\langle 100\rangle$	432	c	1	$\langle$uvw$\rangle$	$\bar{4}$3m	c
2mm1$_R$	$\langle 100\rangle$	m3	c	1	$\langle$uvw$\rangle$	1	a
4mm1$_R$	$\langle 100\rangle$	m3m	c	1	$\langle$uvw$\rangle$	2	m
m1$_R$	$\langle 100\rangle$	mm2	t	1	$\langle$uvw$\rangle$	4	t
2mm1$_R$	$\langle 100\rangle$	mmm	o	1	$\langle$uvw$\rangle$	4mm	t
m1$_R$	$\langle 110\rangle$	$\bar{4}$2m	t	1	$\langle$uvw$\rangle$	23	c
m1$_R$	$\langle 110\rangle$	$\bar{4}$3m	c	1	$\langle$uvw$\rangle$	222	o
2mm1$_R$	$\langle 110\rangle$	4/mmm	t	1	$\langle$uvw$\rangle$	422	t
m1$_R$	$\langle 110\rangle$	4mm	t	1	$\langle$uvw$\rangle$	432	c
2m$_R$m$_R$	$\langle 110\rangle$	422	t	1	$\langle$uvw$\rangle$	m	m
2m$_R$m$_R$	$\langle 110\rangle$	432	c	1	$\langle$uvw$\rangle$	mm2	o
2mm1$_R$	$\langle 110\rangle$	m3m	c	6$_R$	[0001]	$\bar{3}$	r
3m	$\langle 111\rangle$	$\bar{4}$3m	c	6$_R$mm$_R$	[0001]	$\bar{3}$m	r
3	$\langle 111\rangle$	23	c	31$_R$	[0001]	$\bar{6}$	h
3m$_R$	$\langle 111\rangle$	32	r	3m1$_R$	[0001]	$\bar{6}$m2	h
3m$_R$	$\langle 111\rangle$	432	c	3	[0001]	3	r
6$_R$	$\langle 111\rangle$	m3	c	3m	[0001]	3m	r
6$_R$mm$_R$	$\langle 111\rangle$	m3m	c				

Table 10.9. Continued.

D Grp	Zone	Pt Grp	CS	D Grp	Zone	Pt Grp	CS
6	[0001]	6	h	m_R	[uu$\bar{2}\bar{u}$w]	$\bar{6}$m2	h
2_Rmm$_R$	[0001]	6/m	h	m	[uu$\bar{2}\bar{u}$w]	6mm	h
61_R	[0001]	6/m	h	m_R	[uu$\bar{2}\bar{u}$w]	622	h
6mm1_R	[0001]	6/mmm	h	m	[uuw]	$\bar{4}$2m	t
6mm	[0001]	6mm	h	m	[uuw]	4mm	t
6m$_R$m$_R$	[0001]	622	h	m_R	[uuw]	422	t
4_R	[001]	$\bar{4}$	t	m_R	[uv0]	$\bar{4}$	t
4mm1_R	[001]	$\bar{4}$2m	t	m_R	[uv0]	$\bar{4}$2m	t
4_Rmm$_R$	[001]	$\bar{4}$2m	t	m_R	[uv0]	4	t
4	[001]	4	t	2_Rmm$_R$	[uv0]	4/m	t
41_R	[001]	4/m	t	2_Rmm$_R$	[uv0]	4/mmm	t
4mm	[001]	4mm	t	m_R	[uv0]	4mm	t
2m$_R$m$_R$	[001]	222	o	m_R	[uv0]	222	o
4m$_R$m$_R$	[001]	422	t	m_R	[uv0]	422	t
2mm	[001]	mm2	o	m_R	[uv0]	mm2	o
2mm1_R	[001]	mmm	o	2_Rmm$_R$	[uv0]	mmm	o
2	[010]	2	m	m	[uvt0]	$\bar{6}$	h
21_R	[010]	2/m	m	m	[uvt0]	$\bar{6}$m2	h
1_R	[010]	m	m	m_R	[uvt0]	6	h
2_Rmm$_R$	[uū0w]	$\bar{3}$m	h	2_Rmm$_R$	[uvt0]	6/mmm	h
m	[uū0w]	$\bar{6}$m2	h	m_R	[uvt0]	6mm	h
m	[uū0w]	3m	r	m_R	[uvt0]	622	h
2_Rmm$_R$	[uū0w]	6/mmm	h	2_R	[uvtw]	$\bar{3}$	r
m	[uū0w]	6mm	h	2_R	[uvtw]	$\bar{3}$m	r
m_R	[uū0w]	32	r	2_R	[uvtw]	6/m	h
m_R	[u¯u0w]	622	h	2_R	[uvtw]	6/mmm	h
m_R	[u0w]	$\bar{4}$2m	t	2_Rmm$_R$	[uvtw]	6/mmm	h
m_R	[u0w]	2	m	2_R	[uvw]	$\bar{1}$	a
2_Rmm$_R$	[u0w]	2/m	m	2_R	[uvw]	2/m	m
2_Rmm$_R$	[u0w]	4/mmm	t	2_R	[uvw]	4/m	t
m	[u0w]	4mm	t	2_R	[uvw]	4/mmm	t
m_R	[u0w]	222	o	2_Rmm$_R$	[uvw]	4/mmm	t
m_R	[u0w]	422	t	2_R	[uvw]	m3	c
m	[u0w]	m	m	2_R	[uvw]	m3m	c
m	[u0w]	mm2	o	2_R	[uvw]	mmm	o
2_Rmm$_R$	[u0w]	mmm	o				

Table 10.10. Matrix Relating the Diffraction Groups (vertical) to the Point Groups (horizontal) .

BRIGHT FIELD	WHOLE PATTERN	DIFFRACTION GROUPS	RELATION BETWEEN THE DIFFRACTION GROUPS AND THE CRYSTAL POINT GROUPS
3mm	6mm	$6mm1_R$	
6mm	6mm	$3m1_R$	
$\bar{6}mm$	3m	6mm	
3mm	6	$6m_Rm_R$	
6	6	61_R	
6	3	31_R	
6	6	6	
3m	3m	6_Rmm_R	
3m	3m	3m	
3m	3	$3m_R$	
3	3	6_R	
3	3	3	
4mm	4mm	$4mm1_R$	
4mm	2mm	4_Rmm_R	
4mm	4mm	4mm	
4mm	4	$4m_Rm_R$	
4	4	41_R	
4	2	4_R	
4	4	4	
2mm	2mm	$2mm1_R$	
m	m	2_Rmm_R	
2mm	2mm	2mm	
2mm	2	$2m_Rm_R$	
2mm	m	$m1_R$	
m	m	m	
m	1	m_R	
2	2	21_R	
1	1	2_R	
2	2	2	
2	1	1_R	
1	1	1	

Point group columns (horizontal, left to right): 1, $\bar{1}$, 2, m, $2/m$, 222, $mm2$, mmm, 4, $\bar{4}$, $4/m$, 422, $4mm$, $\bar{4}2m$, $4/mmm$, 3, $\bar{3}$, 32, $3m$, $\bar{3}m$, 6, $\bar{6}$, $6/m$, 622, $6mm$, $\bar{6}m2$, $6/mmm$, 23, $m3$, 432, $\bar{4}3m$, $m3m$

Table 10.11. Symmetry Symbols for CBED Discs for the Point Groups.

Table 10.11. Continued.

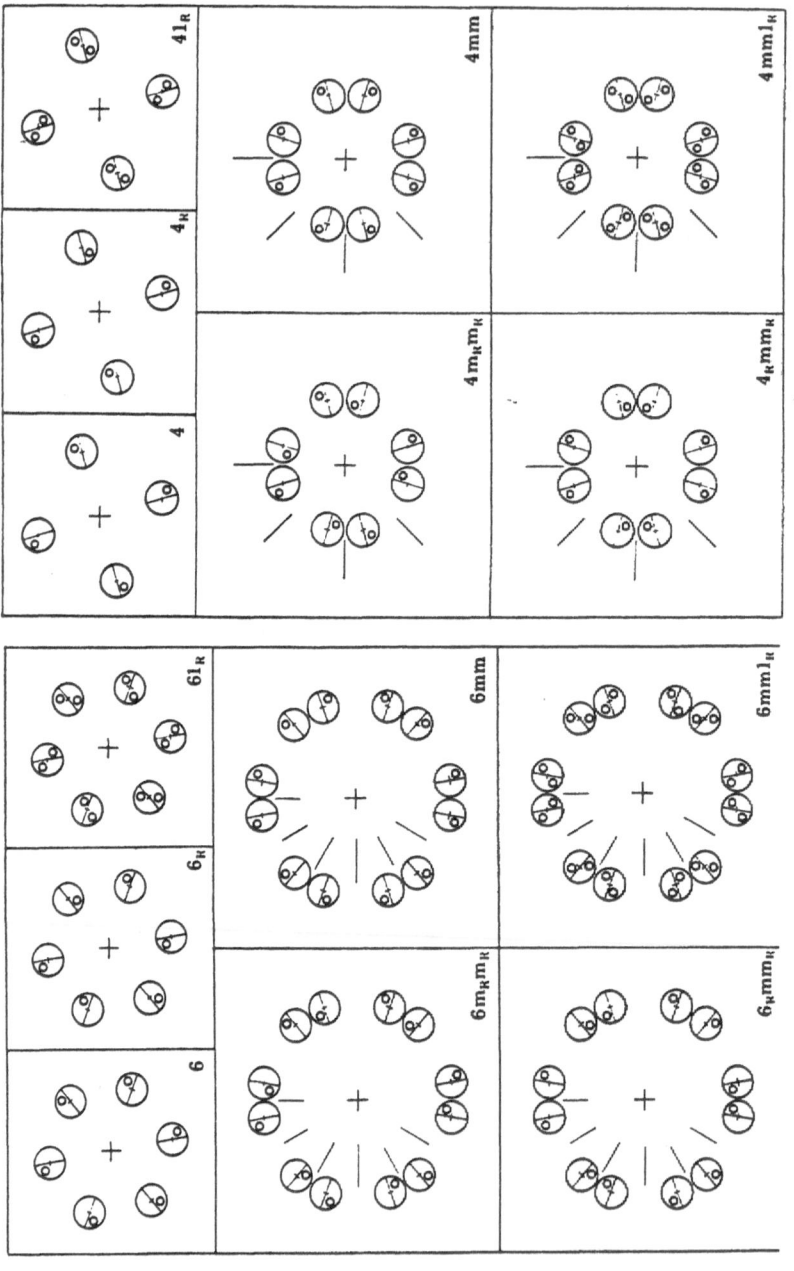

The subscript R refers to a π rotation applied within the disc after the symmetry operation has been performed, e.g., 2_R represents a 2-fold rotation about the symmetry center of the pattern followed by a π rotation about the symmetry axis within the disc. Reprinted from Buxton et al. [1976], by permission of the Royal Society, London.

Table 10.12. Expressions for u* and for H/u* Used in the Calculation of Translation Vectors.

$|u^*|$	$H/|u^*|$

Cubic:

$$\frac{1}{a}\sqrt{u^2 + v^2 + w^2}$$

$$\frac{1}{u^2 + v^2 + w^2}$$

Tetragonal:

$$\frac{1}{a}\sqrt{u^2 + v^2 + (c/a)^2 w^2}$$

$$\frac{1}{u^2 + v^2 + (c/a)^2 w^2}$$

Orthorhombic:

$$\frac{1}{a}\sqrt{u^2 + (b/a)^2 v^2 + (c/a)^2 w^2}$$

$$\frac{1}{u^2 + (b/a)^2 v^2 + (c/a)^2 w^2}$$

Monoclinic[a]:

$$\frac{1}{a}\sqrt{u^2 + (b/a)^2 v^2 + (c/a)^2 w^2 - 2 (c/a) u w \cos \beta^*}$$

$$\frac{1}{u^2 + (b/a)^2 v^2 + (c/a)^2 w^2 - 2 (c/a) u w \cos \beta^*}$$

hcp:

(3,3): $\dfrac{2}{a}\sqrt{u^2 + v^2 - uv + (c/a)^2 w^2}$

$$\frac{1}{2(u^2 + v^2 - uv + (c/a)^2 w^2)}$$

(4,4): $\dfrac{2}{3a}\sqrt{3(u^2 + v^2 + uv) + (c/a)^2 w^2}$

$$\frac{1}{\frac{2}{3}[3(u^2 + v^2 + uv + (c/a)^2 w^2)]}$$

[a] Note: $\beta^* = \pi - \beta$; $90° < \beta < 120°$; b axis unique.

11
Miscellaneous Tables and Data

11.1. Mendeleev Number and Chemical Scale

The usual arrangement of elements in the periodic table makes use of the regularity of the atomic number and atomic weights to categorize the properties of the elements. An alternative categorization is to use the Mendeleev number and chemical scale proposed by Pettifor [1984] and shown in Table 11.1. The Mendeleev number is similar to the atomic number, but the basis for the ordering is the structure the element possesses. The chemical scale is analogous to the electronegativity scale, and the larger the number the more reactive is the element. This categorization is useful for creating structure maps of binary and ternary compounds to determine probable existence of a compound. See Pettifor [1984, 1985, 1986] and Pettifor and Podloucky [1985, 1986] for further details.

11.2. Machlin Classification of Some Intermetallics

Machlin [1974] has categorized groups of compounds using a simple pair potential model. The categories are as follows:

Class	Combination	Notes
Homoelectronic phases		
I	sp-sp	simple or polyvalent metals; no transition metals, rare earth or actinides
II	$\underline{d}$ - ∂	$\underline{d}$ are early TM elements but not CR, Mo, W
		∂ are late TM elements but not Cr, Mo, W
VII	∂ - ∂	
VIII	$\underline{d}$ - $\underline{d}$	
IX	$\underline{d}$ - ldl	ldl denotes Cr, Mo, W
XI	∂ - ldl	
Heteroelectronic phases		
III L	5f - sp	
III A	6f - sp	
IV L	5f - ∂	
IV A	6f - ∂	
V	sp- ∂	
VI	sp - $\underline{d}$	
X	d - $\overline{f}$	

In Table 11.2 the compounds presented in the papers are listed and sorted according to the average of the lattice parameter of the structures, the standard deviation, and the high and the low value of the lattice parameter. The usefulness of the table lies in the clustering of lattice parameters for a given group. In identifying a new phase, the values of the lattice parameters can be compared with those in the Tables to provide additional clues as to the bonding possible or clues as to isostructural phases. The tables, of course, cannot by themselves be used to conclusively identify a phase.

11.3. Schlafli Symbols

These symbols are used to designate the symmetries present in two dimensional tilings of simple polygons [Cundy and Rollett, 1961]. The triangle, square, and hexagon constitute the geometrical elements of the three *regular plane tessellations* (patterns of one kind of congruent regular polygons which fill the plane). The eight *semi-regular tessellations* consist of two or more kinds of polygon arranged in the plane such that every vertex is congruent. The Schlafli symbol (or the modified Schlafli symbol) is generated by noting the types of polygon from which the vertex is generated. For example, in a square net each vertex is generated by the meeting of four squares. Since the square is represented by its 4-fold symmetry (or its four sides) the Schlafli symbol is 4444 or 4^4. A net of regular triangles is 3^6 and a net of regular hexagons is 6^3.

For semi-regular nets, the Schlafli symbol is produced by combining polygon symbols. There are only eight possible semi-regular tilings:

$$3^3\,4^2 \qquad\qquad 3\,12^2$$
$$3^2\,4\,3\,4 \qquad\qquad 4\,8^2$$
$$3\,6\,3\,6 \qquad\qquad 4\,6\,12$$
$$3^4\,6 \qquad\qquad 3\,4\,6\,4$$

The 3^4 6 tiling is the only one that has enantiomorphic forms, which can be distinguished by noting left (L) or right(R) next to the symbol.

The order of the polygon symbols is significant, i.e., 3 6 3 6 is not equivalent to 3^2 6 6 or 3^2 6^2 or 3 3 6^2. The polygon symbols may, however, be permuted, e.g., 3^3 4^2 = 4^2 3^3.

Other tilings are possible using a rhombus or parallelogram, the Penrose tiling being the most recent and well known. The Penrose tile consists of two rhombi, one with internal angles of 72 and 108, and the other with internal angles of 36 and 144. Use of these two rhombi to tile a plane is possible, even though there is no periodicity present in the usual understanding of translational periodicity of a unit cell. The most prominent symmetry present is 5-fold, since the internal angles of a pentagon are 36, 72, and 108, the same as three of the internal angles of the rhombi used [Nelson, 1986].

Combinations of regular and semi-regular nets are possible, and these are written, for example, as 3^3 + 3^4 6. Pentagons are admissible, even though pentagons, by themselves, cannot completely cover the plane.

The regular and semi-regular tilings are shown in Figures 11.1 and 11.2. The Penrose tiling is shown in Figure 11.3.

The relationship among Schlafli symbols, nets, and equivalent points is not obvious, but the connection can be made by arranging the equivalent points of a space group into groups of points that have common z coordinates. Structures of crystals can then be viewed as stacks of layers of nets generated by connecting the atom sites in each layer to form nets of regular, semi-regular, or other polygons. Complications arise in some cases because of the necessity for "rumpled" layers, which are not strictly two dimensional but which display the symmetry of the structure. Detailed application of nets to the description of structures is presented in Pearson's *Crystal Chemistry and Physics of Alloys and Metals* [Pearson, 1972]. Intermetallic structures are particularly well suited to description in terms of nets, and even complex polymer structures or biological structures can be considered in terms of layers rotated about some axis.

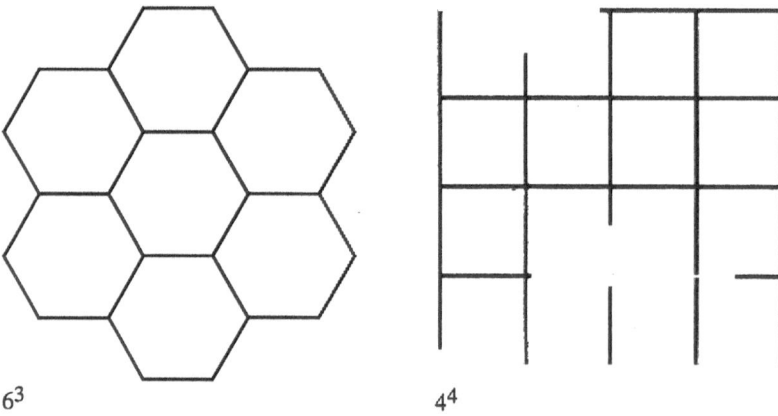

6^3 4^4

3^6

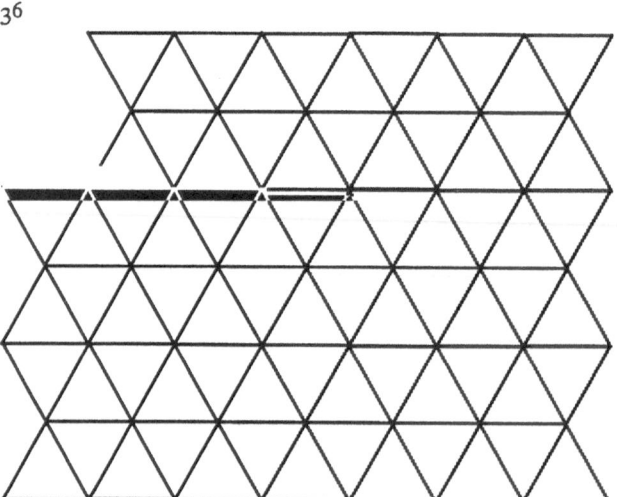

Figure 11.1. The regular tesselations 6^3, 4^4, and 3^6.

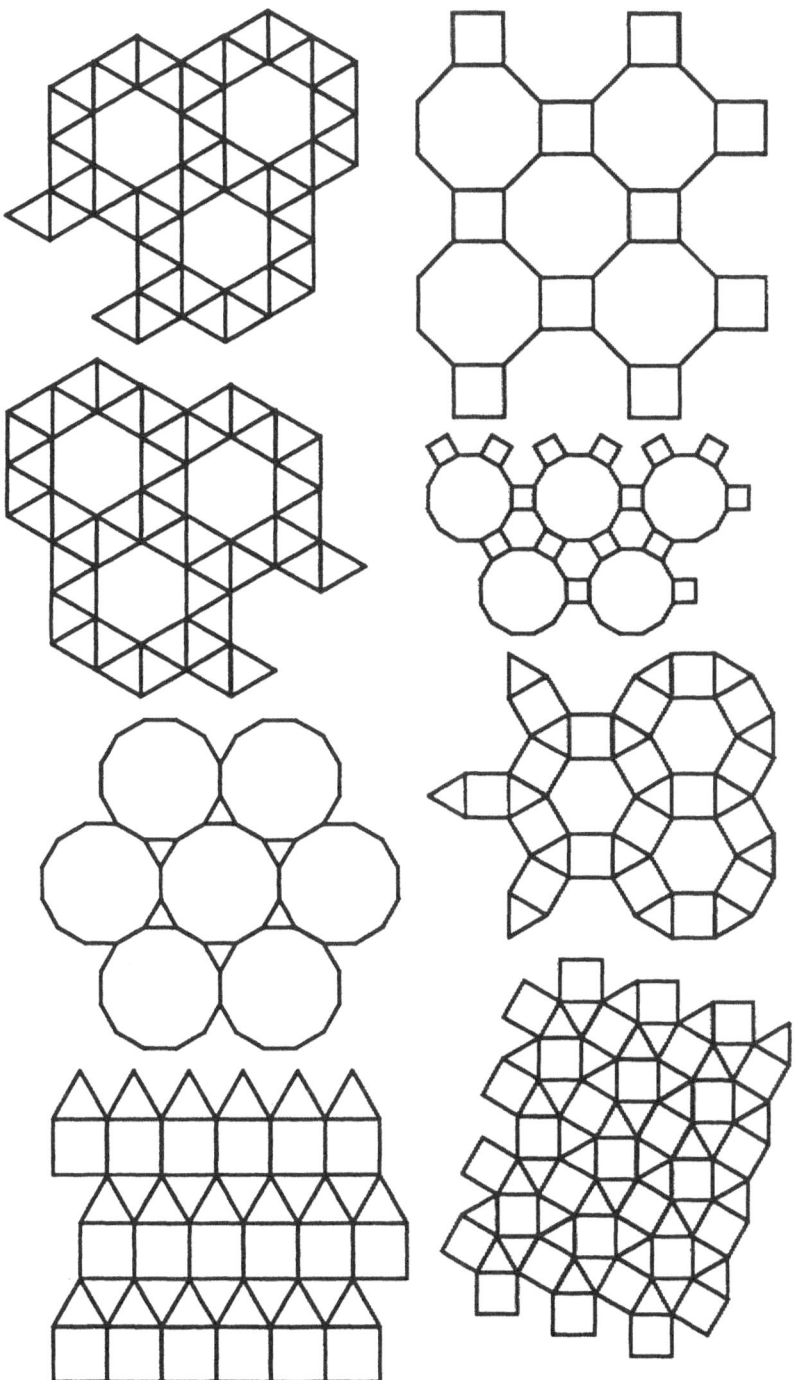

Figure 11.2. The eight semi-regular tesselations: starting in the left column, top down: 3^4 6 - L; 3^4 6 - R: 3 12^2: 3^3 4^2; right column, top down: 4 8^2; 4 6 12; 3 4 6 4; 3^2 4 3 4.

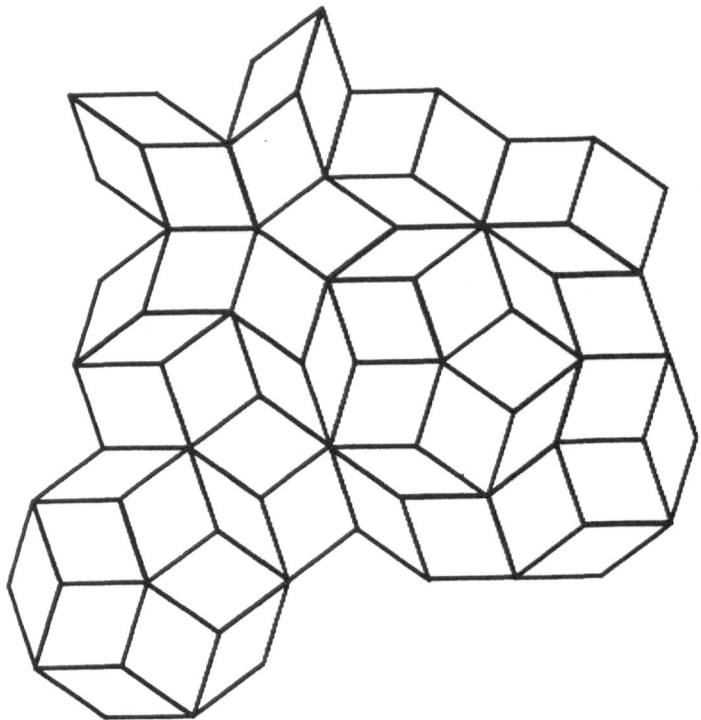

Figure 11.3. The Penrose tile produced by using two rhombi, one with internal angles of 72 and 108 and the other with internal angles of 36 and 144.

11.4. Fourier Series and Transforms

11.4.1. Introduction

Fourier series and Fourier transforms are a convenient way to represent the physical behavior of diffracted waves. Such representations are used extensively in x-ray analyses to obtain electron density data. Electron diffraction takes a somewhat different approach, the Bloch waves approach, but the use of Fourier transforms is still useful for representing the potentials involved in the electron scattering process. In these sections, the basic mathematics of Fourier series and Fourier transforms is listed. Detailed explanations of this subject can be found in [Cowley, 1990], optical applications in [Hecht, 1975] and x-ray applications in [ITXC, 1959].

11.4.2. Fourier series

A function $f(x)$ can be represented as a series made from harmonic functions whose wavelengths are multiples of the wavelength associated with $f(x)$. The expression for the function as a series is

$$f(x) = f_0 + \sum_{n=1}^{\infty} f_n \, e^{2\pi i k x n},$$

$$= f_0 + \sum_{n=1}^{\infty} A_n \cos(nk2\pi x) + i \sum_{n=1}^{\infty} B_n \sin(nk2\pi x), \qquad (11.1)$$

where n is an integer, $k = 1/\lambda$, x is the coordinate. The coefficients A_n and B_n are given by

$$A_n = \frac{2}{\lambda} \int_0^{\lambda} f(x) \cos(nk2\pi x) \, dx, \qquad (11.2)$$

$$B_n = \frac{2}{\lambda} \int_0^{\lambda} f(x) \sin(nk2\pi x) \, dx. \qquad (11.3)$$

If $f(x) = -f(-x)$, i.e., $f(x)$ is an odd function, then $B_n = 0$; if $f(x) = f(-x)$, i.e., $f(x)$ is an even function, then $A_n = 0$.

A simple example is the square wave shown in Figure 11.4, in which the origin has been located at the start of the positive portion of the square wave. The functional expression for this wave is

$$f(x) = \begin{cases} 1 & 0<x<\lambda/2; \ \lambda<x<3\lambda/2; \ ... \\ -1 & \lambda/2<x<\lambda; \ 3\lambda/2<x<4\lambda/2; \ .... \end{cases} \qquad (11.4)$$

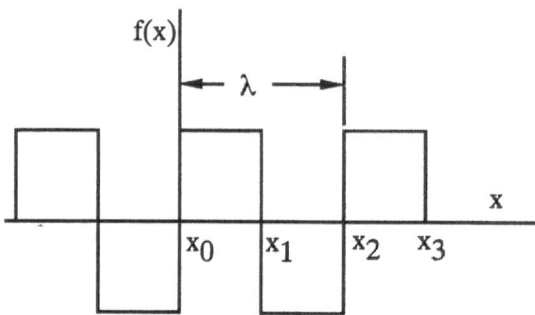

Figure 11.4. Square wave example of Fourier series representation of complicated functions that may be either real or complex.

Because $f(1) = -f(-1)$, the function is odd, and therefore the A_n coefficients are identically zero. The B_n coefficients are found by inserting the appropriate values for $f(x)$ in the definitions and integrating.

Thus,

$$B_n = \frac{2}{\lambda} \int_0^{\lambda/2} (1) \sin (2\pi nkx) \, dx + \frac{2}{\lambda} \int_{\lambda/2}^{\lambda} (-1) \sin (2\pi nkx) \, dx$$

$$= -\frac{[\cos (\pi nk\lambda) - 1]}{\lambda \pi nk}. \tag{11.5}$$

11.4.3. Fourier transforms

If the series representation of function $f(x)$ is generalized to a continuous case, then the series become integrals, written using A_n, B_n and the complex notations as

$$f(x) = \frac{1}{2\pi} \int_{-\infty}^{\infty} F(k) \, e^{-ikx} \, dk,$$

$$= \frac{1}{\pi} \int_0^{\infty} A(k) \cos (kx) \, dx + \frac{i}{\pi} \int_0^{\infty} B(k) \sin (kx) \, dx. \tag{11.6}$$

The function $F(k)$ is defined as

$$F(k) = \int_{-\infty}^{\infty} f(x) \, e^{ikx} \, dx = A(k) + i B(k), \tag{11.7}$$

which is *defined* to be the *Fourier transform of the function f(x)*.

Extension of the Fourier series and transform to two and three dimensions is direct; $f(x)$ is replaced by $f(x,y)$ or $f(x,y,z)$ and $F(k)$ by $F(k_x,k_y)$ and $F(k_x, k_y, k_z)$, and the integrals are taken over the appropriate dimension.

Dirac delta function: A special function useful for representing pulses or very sharp events is the Dirac delta function, δ:

$$\delta(x) = \begin{cases} 0 & x \neq 0 \\ \infty & x = 0, \end{cases} \tag{11.8a}$$

$$\int_{-\infty}^{\infty} \delta(x)\, dx = 1. \tag{11.8b}$$

It is also referred to as the unit impulse function.

The delta function also can be used as a filter:

$$\int_{-\infty}^{\infty} \delta(x - x_0)\, f(x)\, dx = f(x_0). \tag{11.9}$$

The Fourier transform of the delta function:

$$F(k) = e^{ikx_0}. \tag{11.10}$$

Convolution: The convolution of two functions $f(x)$ and $g(x)$ is defined, with $*$ as the operator symbol, as

$$f(x) * g(x) = \int_{-\infty}^{\infty} f(x')\, g(x - x')\, dx'$$

$$= g(x) * f(x); \tag{11.11}$$

in more general form using vectors,

$$f(x) * g(x) = \int_{-\infty}^{\infty} f(x')\, g(x - x')\, dx'$$

$$= g(x) * f(x). \tag{11.12}$$

Table 11.3 gives some common transforms.

11.4.4. Crystal structures and Fourier transforms

In x-ray diffraction, the electron density is represented by a function $\rho(xyz)$. The scattering produced by this distribution of electrons is represented by the function $f(g_x,g_y,g_z)$. The functions are the Fourier transforms of each other, i.e.,

$$\rho(xyz) = \frac{1}{V} \int_{-\infty}^{\infty} \int_{-\infty}^{\infty} \int_{-\infty}^{\infty} f(g_x,g_y,g_z) \exp\left[-2\pi i(g_x x + g_y y + g_z z)\right] dg_x \, dg_y \, dg_z, \qquad (11.13)$$

and the variables are defined by some position vector in direct space, $\mathbf{r}$,

$$\mathbf{r} = x\mathbf{a} + y\mathbf{b} + z\mathbf{c}, \qquad (11.14)$$

and a position vector in reciprocal space, $\mathbf{g}$,

$$\mathbf{g} = g_x\mathbf{a}^* + g_y\mathbf{b}^* + g_z\mathbf{c}^*. \qquad (11.15)$$

The function $f(g_x,g_y,g_z)$ is the scattering function and is the Fourier transform of the density function:

$$f(g_x,g_y,g_z) = V \int_{-\infty}^{\infty} \int_{-\infty}^{\infty} \int_{-\infty}^{\infty} \rho(xyz) \exp\left[2\pi i(g_x x + g_y y + g_z z)\right] dx \, dy \, dz. \quad (11.16)$$

For electron diffraction the density function is replaced by the crystal potential, $U(xyz)$. Its Fourier transform is the electron scattering function from the potentials of the crystal array.

Table 11.1. Mendeleev Number (M) and Chemical Scale (χ).

M	El	χ	M	El	χ	M	El	χ
1	He	0.00	36	Md	0.7125	71	Ag	1.18
2	Ne	0.04	37	Fm	0.715	72	Cu	1.20
3	A	0.08	38	Es	0.7175	73	Mg	1.28
4	Kr	0.12	39	Cf	0.72	74	Hg	1.32
5	Xe	0.16	40	Bk	0.7225	75	Cd	1.36
6	Rñ	0.20	41	Cm	0.725	76	Zn	1.44
7	Fr	0.23	42	Am	0.7275	77	Be	1.50
8	Cs	0.25	43	Pu	0.73	78	Tl	1.56
9	Rb	0.30	44	Np	0.7325	79	In	1.60
10	K	0.35	45	U	0.735	80	Al	1.66
11	Na	0.4	46	Pa	0.7375	81	Ga	1.68
12	Li	0.45	47	Th	0.74	82	Pb	1.80
13	Ra	0.48	48	Ac	0.7425	83	Sn	1.84
14	Ba	0.50	49	Zr	0.76	84	Ge	1.90
15	Sr	0.55	50	Hf	0.775	85	Si	1.94
16	Ca	0.60	51	Ti	0.79	86	B	2.00
17	Yb	0.645	52	Nb	0.82	87	Bi	2.04
18	Eu	0.655	53	Ta	0.83	88	Sb	2.08
19	Y	0.66	54	V	0.84	89	As	2.16
20	Sc	0.67	55	Mo	0.88	90	P	2.18
21	Lu	0.675	56	W	0.885	91	Po	2.28
22	Tm	0.6775	57	Cr	0.89	92	Te	2.32
23	Er	0.68	58	Tc	0.935	93	Se	2.40
24	Ho	0.6825	59	Re	0.94	94	S	2.44
25	Dy	0.685	60	Mn	0.945	95	C	2.50
26	Tb	0.6875	61	Fe	0.99	96	At	2.52
27	Gd	0.69	62	Os	0.995	97	I	2.56
28	Sm	0.6925	63	Ru	1.00	98	Br	2.64
29	Pm	0.695	64	Co	1.04	99	Cl	2.70
30	Nd	0.6975	65	Ir	1.05	100	N	3.00
31	Pr	0.70	66	Rh	1.06	101	O	3.50
32	Ce	0.7025	67	Ni	1.09	102	F	4.00
33	La	0.705	68	Pt	1.105	103	H	5.00
34	Lw	0.7075	69	Pd	1.12			
35	No	0.71	70	Au	1.16			

Based on Pettifor [1984].

Table 11.2. Lattice Parameters of Some Intermetallic Compounds by Strukturbericht Symbol and Machlin Classification.

Symb	<>	d	High	Low	Symb	<>	d	High	Low
C15 I	7.179	0.265	4.032	2.703	C15 VIII	7.335	0.125	7.445	7.160
C15 II	6.965	0.237	7.359	6.918	C15b I	6.095	-	-	-
$L1_2$ I	4.358	0.581	4.901	3.743	C15b II	6.700	0.021	6.718	6.689
$L1_2$ II	3.89	0.054	3.981	3.795	C15b III	7.029	-	-	-
A15 II	4.927	0.182	5.153	4.675	C15b IV	6.678	-	-	-
C15b I	6.095	-	-	-	C15b V	5.940	0.080	5.996	5.883
C15b II	6.7	0.021	6.718	6.689	$L1_2$ I	4.358	0.581	4.901	3.743
B2 III	3.783	0.115	3.963	3.502	$L1_2$ II	3.890	0.054	3.981	3.795
C15 III	8.118	0.361	8.787	7.76	$L1_2$ III	4.855	0.122	5.093	4.714
$L1_2$ III	4.855	0.122	5.093	4.714	$L1_2$ IV	4.122	0.088	4.235	4.027
C15b III	7.029	-	-	-	$L1_2$ V	3.891	0.153	4.053	3.682
B2 IV	7.457	0.263	7.774	6.624	$L1_2$ VI	4.153	0.218	4.365	3.930
$L1_2$ IV	4.122	0.088	4.235	4.027	$L1_2$ VII	3.735	0.136	3.831	3.639
C15b IV	6.678	-	-	-					

Symb	<>	d	High	Low
B2 V	2.933	0.158	3.166	2.88
C15 V	7.727	0.143	7.953	7.525
$L1_2$ V	3.891	0.153	4.053	3.682
C15b V	5.94	0.080	5.996	5.883
B2 VI	3.509	0.168	3.8	3.256
C15 VI	7.052	0.629	7.855	6.44
$L1_2$ VI	4.153	0.218	4.365	3.93
A15 VI	5.151	0.152	5.289	4.88
$L1_2$ VII	3.735	0.136	3.831	3.639
C15 VIII	7.335	0.125	7.445	7.16
C15 IX	7.28	0.318	7.615	6.94
A15 XI	4.973	0	4.973	4.973
B2 I	3.437	0.393	4.032	2.703
B2 II	3.166	0.0948	3.283	2.991

Arranged by average lattice parameter:

Symb	<>	d	High	Low
B2 V	2.933	0.158	3.166	2.880
B2 II	3.166	0.095	3.283	2.991
B2 I	3.437	0.393	4.032	2.703
B2 VI	3.509	0.168	3.800	3.256
$L1_2$ VII	3.735	0.136	3.831	3.639
B2 III	3.783	0.115	3.963	3.502
$L1_2$ II	3.890	0.054	3.981	3.795
$L1_2$ V	3.891	0.153	4.053	3.682
$L1_2$ IV	4.122	0.088	4.235	4.027
$L1_2$ VI	4.153	0.218	4.365	3.930
$L1_2$ I	4.358	0.581	4.901	3.743
$L1_2$ III	4.855	0.122	5.093	4.714
A15 II	4.927	0.182	5.153	4.675
A15 XI	4.973	0.000	4.973	4.973
A15 VI	5.151	0.152	5.289	4.880
C15b V	5.940	0.080	5.996	5.883
C15b I	6.095	-	-	-
C15b IV	6.678	-	-	-
C15b II	6.700	0.021	6.718	6.689
C15 II	6.965	0.237	7.359	6.918
C15b III	7.029	-	-	-
C15 VI	7.052	0.629	7.855	6.440
C15 I	7.179	0.265	4.032	2.703
C15 IX	7.280	0.318	7.615	6.940
C15 VIII	7.335	0.125	7.445	7.160
B2 IV	7.457	0.263	7.774	6.624
C15 V	7.727	0.143	7.953	7.525
C15 III	8.118	0.361	8.787	7.760

Arranged by symbol:

Symb	<>	d	High	Low
A15 II	4.927	0.182	5.153	4.675
A15 VI	5.151	0.152	5.289	4.880
A15 XI	4.973	0.000	4.973	4.973
B2 I	3.437	0.393	4.032	2.703
B2 II	3.166	0.095	3.283	2.991
B2 III	3.783	0.115	3.963	3.502
B2 IV	7.457	0.263	7.774	6.624
B2 V	2.933	0.158	3.166	2.880
B2 VI	3.509	0.168	3.800	3.256
C15 I	7.179	0.265	4.032	2.703
C15 II	6.965	0.237	7.359	6.918
C15 III	8.118	0.361	8.787	7.760
C15 IX	7.280	0.318	7.615	6.940
C15 V	7.727	0.143	7.953	7.525
C15 VI	7.052	0.629	7.855	6.440

Table 11.2. Continued.

Symb	<>	d	High	Low	Symb	<>	d	High	Low
Arranged by largest range of lattice parameter:					$L1_2$ IV	4.122	0.088	4.235	4.027
B2 V	2.933	0.158	3.166	2.880	A15 II	4.927	0.182	5.153	4.675
B2 II	3.166	0.095	3.283	2.991					
B2 VI	3.509	0.168	3.800	3.256	$L1_2$ III	4.855	0.122	5.093	4.714
$L1_2$ VII	3.735	0.136	3.831	3.639	A15 VI	5.151	0.152	5.289	4.880
B2 III	3.783	0.115	3.963	3.502	A15 XI	4.973	0.000	4.973	4.973
$L1_2$ II	3.890	0.054	3.981	3.795	C15b V	5.940	0.080	5.996	5.883
C15 I	7.179	0.265	4.032	2.703	C15 VI	7.052	0.629	7.855	6.440
B2 I	3.437	0.393	4.032	2.703	B2 IV	7.457	0.263	7.774	6.624
$L1_2$ V	3.891	0.153	4.053	3.682	C15b II	6.700	0.021	6.718	6.689
$L1_2$ IV	4.122	0.088	4.235	4.027	C15 II	6.965	0.237	7.359	6.918
$L1_2$ VI	4.153	0.218	4.365	3.930	C15 IX	7.280	0.318	7.615	6.940
$L1_2$ I	4.358	0.581	4.901	3.743	C15 VIII	7.335	0.125	7.445	7.160
A15 XI	4.973	0.000	4.973	4.973	C15 V	7.727	0.143	7.953	7.525
$L1_2$ III	4.855	0.122	5.093	4.714	C15 III	8.118	0.361	8.787	7.760
A15 II	4.927	0.182	5.153	4.675	C15b I	6.095	-	-	-
A15 VI	5.151	0.152	5.289	4.880	C15b IV	6.678	-	-	-
C15b V	5.940	0.080	5.996	5.883	C15b III	7.029	-	-	-
C15b II	6.700	0.021	6.718	6.689					
C15 II	6.965	0.237	7.359	6.918					
C15 VIII	7.335	0.125	7.445	7.160					
C15 IX	7.280	0.318	7.615	6.940					
B2 IV	7.457	0.263	7.774	6.624					
C15 VI	7.052	0.629	7.855	6.440					
C15 V	7.727	0.143	7.953	7.525					
C15 III	8.118	0.361	8.787	7.760					
C15b I	6.095	-	-	-					
C15b IV	6.678	-	-	-					
C15b III	7.029	-	-	-					

Arranged by smallest range of lattice parameter:				
C15 I	7.179	0.265	4.032	2.703
B2 I	3.437	0.393	4.032	2.703
B2 V	2.933	0.158	3.166	2.880
B2 II	3.166	0.095	3.283	2.991
B2 VI	3.509	0.168	3.800	3.256
B2 III	3.783	0.115	3.963	3.502
$L1_2$ VII	3.735	0.136	3.831	3.639
$L1_2$ V	3.891	0.153	4.053	3.682
$L1_2$ I	4.358	0.581	4.901	3.743
$L1_2$ II	3.890	0.054	3.981	3.795
$L1_2$ VI	4.153	0.218	4.365	3.930

Based on data in Machlin [1974].

Table 11.3. Some Common Fourier Transforms.

Function f(x)	Transform F(k)

One-dimensional slit:

$$f(x) = \begin{matrix} 1 \text{ if } |x|<a/2 \\ 0 \text{ if } |x|>a/2 \end{matrix} \qquad F(k) = a\frac{\sin(\pi ak)}{\pi ak}$$

Point source (a → 0):

$$f(x) = \delta(x) \qquad F(k) = 1$$

$$f(x) = \delta(x - b) \qquad F(k) = \exp(2\pi ibk)$$

Transparent rectangle:

$$f(x, y) = \begin{matrix} a - a/2<x<a/2 \\ b - b/2<y<b/2 \end{matrix} \qquad F(k) = ab\frac{\sin(\pi ak_x)}{\pi ak_x}\frac{\sin(\pi ak_y)}{\pi ak_y}$$

Parallelepiped:

$$f(x,y,z) = f(r) = \begin{matrix} a - a/2<x<a/2 \\ b - b/2<y<b/2 \\ c - c/2<z<c/2 \end{matrix} \qquad F(k) = abc\frac{\sin(\pi ak_x)}{\pi ak_x}\frac{\sin(\pi ak_y)}{\pi ak_y}\frac{\sin(\pi ck_z)}{\pi ck_z}$$

Circular hole:

$$f(r) = \begin{matrix} 1 \text{ if } |r|<R \\ 0 \text{ if } |r|<R \end{matrix} \quad F(k) = \pi R^2\frac{J_1(2\pi kR)}{2\pi kR} \; ; J_1 \text{ is the spherical Bessel function.}$$

Sphere of radius R:

$$f(r) = \frac{4\pi r^3}{3} \qquad F(k) = \frac{4\pi R^3}{3} 3\frac{\sin(2\pi kR) - 2\pi kR\cos(2\pi kR)}{(2\pi kR)^3}$$

Gaussian function:

$$f(x) = \exp\left[-\left(\frac{x}{a}\right)^2\right] \qquad F(k) = \sqrt{\pi}\, a\exp\left[-(\pi ka)^2\right]$$

One-dimensional point lattice, with spacing between lattice points of d, N the number of points:

$$f(x) = \sum_{n}^{N} \delta(x - x_n) \qquad F(k) = \frac{\sin(\pi kNd)}{\sin(\pi kd)}$$

Three-dimensional point lattice:

$$f(r) = \sum_{u=-\infty}^{+\infty} \sum_v \sum_w \delta(r - (ua_1 + va_2 + wa_3))$$

$$F(k) = \sum_h \sum_k \sum_l \delta(k - (ha^*_1 + ka^*_2 + la^*_3))$$

For additional examples see Reimer [1988, p.72f.]. Table based on Reimer [1988].

12
Icosahedral Structures and Patterns

12.1. Definitions

12.1.1. Golden mean

In terms of ratios of segments of a line, the golden mean (also golden ratio or golden section) is defined as the ratios of the section illustrated in Figure 12.1:

$$\frac{CB}{AC} = \frac{AC}{AB}. \qquad (12.1)$$

Numerically the value of the golden mean is 1.618033... . In terms of ratios of the side of a rectangle with one side 1 unit and the second side 2 units, the golden mean is

$$\tau = \frac{1 + \sqrt{5}}{2}, \qquad (12.2)$$

as shown in Figure 12.1(b). Hence, the golden mean is a simple way to estimate the length of the hypotenuse if the side ratios are 2:1.

Fibonacci numbers are defined as 1,2,3,5,8,13,..., i.e., the next adjacent number is the sum of the previous two numbers. The golden mean is defined exactly as the limit to the sequence of ratios of adjacent Fibonacci numbers:

$$\tau = \lim_{n \to \infty} \{ 2/1, 3/2, 5/3, 8/5, 13/8, ... \} = 1.618033... . \qquad (12.3)$$

In terms of a circle and an inscribed decagon, the golden mean is the ratio of the radius to the secant length of the inscribed decagon, as shown in Figure 12.1(c).

12.1.2. Icosahedron

An icosahedron, Figure 12.2, is defined as a 20-sided object resulting when 20 tetrahedra are close packed [Cundy and Rollett, 1961; Nelson, 1986]. The coordination number is approximately 12; the icosahedron exhibits quasiperiodicity in three dimensions.

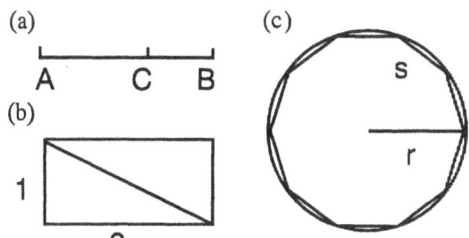

Figure 12.1. Golden mean defined in terms (a) of ratios of line segments, (b) of the geometry of a simple rectangle, (c) of the ratio of the radius of a circle to the secant length of an inscribed decagon. $\tau = 1.618033... .$

12.2. Axes

Because of the geometrical complexity of the icosahedron, defining axes suitable for interpretation of electron diffraction patterns becomes a difficult task. Several schemes were originally used, but the literature appears to be settling on indices described mathematically by [Elser, 1985, 1986]. These Elser indices consist of integers associated with six vectors which allow the 5-fold symmetries to be explained. Other symmetries normally forbidden by classical symmetry are also encompassed in Elser's approach.

The basis set of vectors consists of six vectors aligned along the 5-fold axes of the icosahedron. The origin of the vectors is placed at the vertex as shown in Figure 12.3, and the vectors are numbered in a counterclockwise sequence (right-handed system), the first vector being defined to be the polar vector. A vector in the direct lattice, thus, is given as

$$\mathbf{u} = \sum_{n=1}^{6} u_i \, \mathbf{e}_i \frac{a_r}{\pi},$$ (12.4)

where $\mathbf{e}_i$ are unit vectors, a_r is defined as the edge length of the rhombohedral cells in the three-dimensional Penrose tile, and u_i is an integer.

The reciprocal lattice is defined using

$$\mathbf{g} = \sum_{i=1}^{6} n_i \, \mathbf{e}^{*}_i \frac{\pi}{a_r},$$ (12.5)

where $\mathbf{e}^{*}_i$ are reciprocal unit vectors derived in the usual way from $\mathbf{e}_i$, and n_i is an integer.

Other indices have been used in which a_r is chosen differently but in which six indices are used. The method (introduced by Bancel et al. [1985]) uses twelve vectors generated by cyclic combinations of $(\pm 1, \pm\tau, 0)$. There are six independent vectors, the other six being linear combinations of these. Another system uses vectors referred to cubic basis vectors, with the icosahedral basis being defined by $(\pm\tau, 0, \pm\tau)$ [Cahn et al., 1986].

From Elser's formulation, interpretation of the electron diffraction pattern is accomplished by assuming a basis set of vectors to be

$$\text{basis} = \{\mathbf{e}^i_{\parallel} , i = 1, ... , p\}$$ (12.6)

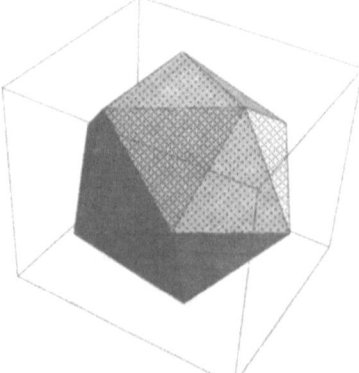

Figure 12.2. The icosahedron. Close packing of 20 tetrahedra produces this geometrical solid.

All peaks can be expressed as a linear combination of these basis vectors or as integral multiples of the basis vectors. The vector $e^i_{\parallel}$ is in three dimensions and is part of the six-dimensional vector

$$e^i = (e^i_{\parallel}, e^i_{\perp}). \tag{12.7}$$

Orient the $e^i_{\parallel}$ vectors along the six 5-fold symmetry axes of the icosahedron. The convention for orienting the six vectors in the projection is shown in Figure 12.3. The polar vector is $e^1_{\parallel}$, and $e^2_{\parallel}$ through $e^6_{\parallel}$ are arranged counterclockwise about the polar vector in a right-handed system. The $e^i_{\perp}$ are shown in the Figure in the configuration which results from the convention choice for the six $e^i_{\parallel}$ vectors.

The dot products are

$$e^i_{\parallel} \cdot e^j_{\parallel} = -e^i_{\perp} \cdot e^j_{\perp} = \pm \frac{1}{\sqrt{5}}, \, (i \neq j), \tag{12.8}$$

where the + is used with the parallel and the − with the perpendicular vectors.

Reciprocal vectors are defined as

$$g_{\parallel} = \frac{\pi}{a} \sum_{i=1}^{6} n_i e^i_{\parallel}, \tag{12.9}$$

where n_i are integers and are the diffracted spot indices, a is the quasilattice constant equal to the edge length of the rhombohedral cells in the three-dimensional Penrose tiling. The corresponding perpendicular g vector is defined as

$$g_{\perp} = \frac{\pi}{a} \sum_{i=1}^{6} n_i e^i_{\perp}. \tag{12.10}$$

The Fourier transform of a density ρ is a function of $g_{\perp}$ and is the structure factor. The density is constant in a region T, which is equivalent to a triacontahedron, a 30-sided solid.

Indexing is done via "inflation" and "deflation." Inflation is done by using the matrix

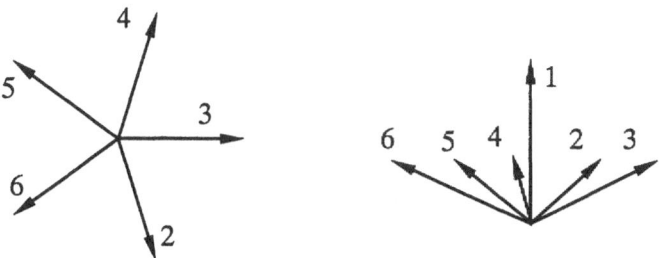

Figure 12.3. Convention used for orienting the six 5-fold axes of the icosahedron.

$$M_{ij} = \frac{1}{2} \begin{vmatrix} 1 & 1 & 1 & 1 & 1 & 1 \\ 1 & 1 & 1 & -1 & -1 & 1 \\ 1 & 1 & 1 & 1 & -1 & -1 \\ 1 & -1 & 1 & 1 & 1 & -1 \\ 1 & -1 & -1 & 1 & 1 & 1 \\ 1 & 1 & -1 & -1 & 1 & 1 \end{vmatrix}_{ij} , \qquad (12.11)$$

and noting that

$$M^2 = M + 1, \qquad (12.12)$$

and that the eigenvalues are the golden mean and the negative reciprocal, i.e., τ and $-1/\tau$. Deflation is the opposite of inflation and

$$M^{-1} = M - 1. \qquad (12.13)$$

The expressions for inflated and deflated vectors using the M matrix are

$$(e^i_\parallel)' = \sum_j M_{ij}\, e^j_\parallel = \tau\, e^i_\parallel \quad \text{(inflated by } \tau\text{)}, \qquad (12.14)$$

$$(e^i_\perp)' = \sum_j M_{ij}\, e^j_\perp = -\tau\, e^i_\perp \quad \text{(deflated by } 1/\tau \text{ and reversed)}. \qquad (12.15)$$

Integer indices of a pair of peaks:

$$n'_i = \sum_j M_{ij}\, n_j , \qquad (12.16)$$

for $\sum\limits_{i=1}^{6} n_i$ = even integer, and

$$n'_i = \sum_j (2\, M_{ij} +1)\, n_j, \qquad (12.17)$$

for $\sum\limits_{i=1}^{6} n_i$ = odd integer.

Systematic peaks are separated in reciprocal space according to the rules

$$\tau^k\, g_\parallel , \; (-\tau^{-1})^k\, g_\perp \text{ for even parity}, \qquad (12.18)$$

$$\tau^{3k}\, g_\parallel , \; (-\tau^{-3})^k\, g_\perp \text{ for odd parity}, \qquad (12.19)$$

for $k = 0, \pm1, \pm2, \dots$

The magnitude of the reciprocal lattice vectors is given for the first few vectors in Table 12.1 [see Elser, 1985]. The method of calculation is analogous to the method used in ordinary three dimensional vector calculations. To find the

Table 12.1. A Listing of Peaks for the Icosahedral Quasicrystal Having $|n_{||}|<25$ and $n_\perp<3.4$.

$|n_{||}|$	$|n_\perp|$	$i(n_\perp)$	Parity	n_1	n_2	n_3	n_4	n_5	n_6
0.000	0.000	1.000 000	0	0	0	0	0	0	0
3.142	3.142	0.000 043	1	1	0	0	0	0	0
5.345	3.303	0.000 800	0	1	1	0	0	0	0
7.489	1.768	0.238 225	1	1	1	1	0	0	0
8.469	2.042	0.134 874	0	1	1	1	1	0	0
11.440	2.701	0.013 211	1	2	1	1	1	0	0
12.230	2.887	0.003 335	0	2	1	1	1	1	0
13.308	0.742	0.794 142	1	2	1	1	1	1	1
13.993	1.262	0.501 572	0	2	2	1	0	0	1
14.341	3.385	0.001 817	1	2	2	1	1	0	1
15.871	2.172	0.096 895	1	2	2	2	1	0	0
16.450	2.400	0.047 775	0	3	1	1	1	1	1
18.074	2.981	0.001 404	1	3	2	1	1	1	1
18.584	3.151	0.000 001	0	3	2	2	1	0	0
19.311	1.464	0.387 990	1	3	2	2	1	0	1
19.789	1.784	0.230 827	0	3	2	2	1	1	1
21.159	2.512	0.031 390	1	3	3	2	0	0	1
21.596	2.711	0.012 579	0	3	3	2	1	0	1
22.641	0.780	0.774 703	0	3	3	2	0	0	2
22.858	3.237	0.000 253	1	4	2	2	1	1	1
23.263	3.394	0.001 633	0	3	3	3	1	0	0
23.847	1.932	0.172 020	1	3	3	3	1	0	1
24.236	2.185	0.093 199	0	4	2	2	2	1	1

The intensities given correspond to the primitive quasicrystal with identical point charges at the rhombohedral vertices. The values of $|g_{||}|$ for $Al_{0.86}Mn_{0.14}$ may be obtained by dividing the entries in the first column by $a = 4.60Å$. Reprinted from Elser [1985] by permission.

magnitude, for example of the vector with indices (10 00 00), the definitions of the dot products for the basis vectors are used. Thus

$$|g_{||}(10\ 00\ 00)| = p\sqrt{e_{||}^1 \cdot e_{||}^1} = \pi; \tag{12.20a}$$

$$|g_{||}(11\ 00\ 00)| = p\sqrt{e_{||}^1 \cdot e_{||}^1 + 2 e_{||}^1 \cdot e_{||}^2 + e_{||}^2 \cdot e_{||}^2}$$
$$= \pi\sqrt{1 + \frac{2}{\sqrt5} + 1}$$
$$= \pi\ 1.7013 = 5.345; \tag{12.20b}$$

$$|g_\perp(10\ 00\ 00)| = \pi; \tag{12.20c}$$

$$|g_\perp(11\ 00\ 00)| = \pi\sqrt{1 - \frac{2}{\sqrt5} + 1} = 3.303. \tag{12.20d}$$

12.3. Simple Projection Examples

An example illustrates the method of -dimensional projection onto three dimensions. In Figures 12.4 and 12.5 is a simple square lattice of points [Katz and Duneau, 1986]. A line drawn at some angle with respect to one of the axes (horizontal in Figure 12.5) cuts across the lattice, intercepting some points and missing others. Movement in the lattice is restricted to occur only from point to point by vertical or horizontal moves. If one starts at the origin and moves to the right so that the line is never crossed, the resulting sawtooth pattern can be projected onto the line by projecting each intercepted point on the sawtooth curve. The length of the projections are of two kinds. the first is related to the projection of the vertical distance in the lattice, and the second is the projection of the horizontal length. The distances on the line are labeled a and b. The projection is the "tile" which consists of two tiles a and b. If the ratio of b to a is the golden mean, then the lattice is icosahedral. Note that the slope of the line is irrational.

From Elser [1985], we can define vectors $x = (x_{\parallel}, x_{\perp})$ and $g = (g_{\parallel}, g_{\perp})$. From the properties of reciprocal vectors,

$$\exp (i\, x \cdot g) = \exp (i\, x_{\parallel}\, g_{\parallel}) \exp (i\, x_{\perp}\, g_{\perp}) = 1. \tag{12.21}$$

For N projections of $x^n{}_{\parallel}$, the one-dimensional structure factor is

$$S(g_{\parallel}) = \sum_{n=1}^{N} \exp (i\, x^n{}_{\parallel}\, g_{\parallel}) = \sum_{n=1}^{N} \exp (-\,i\, x^n{}_{\perp}\, g_{\perp}). \tag{12.22}$$

The Fourier transform of this structure factor is

$$S(g_{\parallel}) = \frac{N \sin z}{z}, \tag{12.23}$$

where

$$z = \frac{1}{2}|a - b|\, g_{\perp}, \tag{12.24}$$

for (a,b) $\geq$ interval in which the projection is defined and a and b lie parallel to $x_{\perp}$.

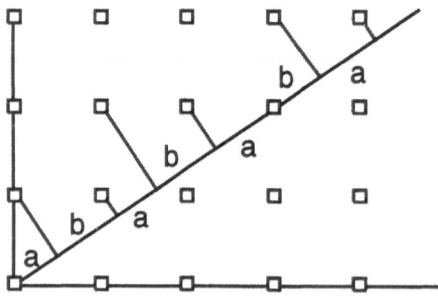

Figure 12.4. Projection onto a line to produce the line segments a and b. b/a is the golden mean for icosahedral patterns. Slope of the line is irrational.

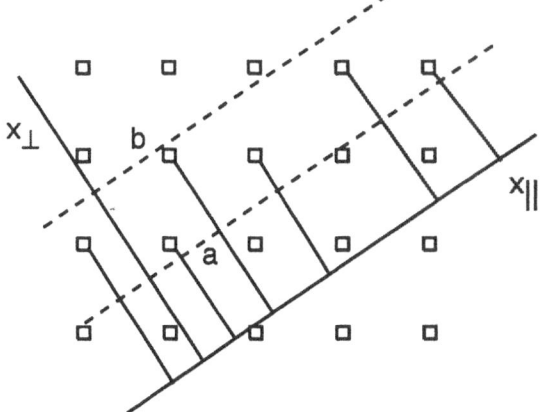

Figure 12.5. One-dimensional projection definitions of $x_\perp$ and $x_\parallel$ and a, b. [Elser, 1985]. Only points lying between a and b are projected.

12.4. Diffraction Patterns

The diffraction patterns from icosahedral structures have specific geometric properties and intensity behavior which make them unique. The separation of diffracted beams in a normal pattern is regular, the separation being defined as the reciprocal lattice vector. In an icosahedral pattern, the separation of diffracted beams along a given vector from the origin is not uniform. The distances are related by the golden ratio, as indicated in Figure 12.6. Analytically this can be written simply as

$$B = \tau A, \ C = \tau B = \tau^2 A = 2.618A,$$
$$D = \tau C = \tau^3 A = 4.236A,$$
$$E = \tau D = \tau^4 A = 6.854A,$$
$$F = \tau E = \tau^5 A = 11.090A,$$
$$G = \tau F = \tau^6 A = 17.944A, \tag{12.25}$$
$$\vdots$$

and the ratios of adjacent lengths for A =1 is

$$\frac{B}{A} = \frac{C}{B} = \frac{D}{C} = \frac{E}{D} = ... = \tau. \tag{12.26}$$

The intensities in a normal diffraction pattern decrease with increasing distance from

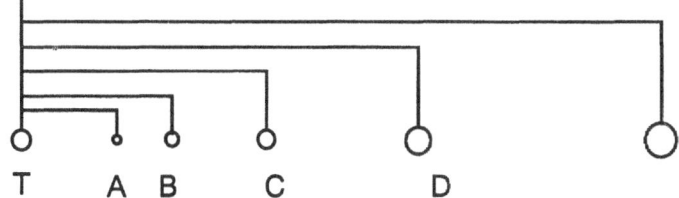

Figure 12.6. Illustration of relative lengths between a row of diffracted beams in an icosahedral pattern.

The intensities in a normal diffraction pattern decrease with increasing distance from the transmitted beam. In an icosahedral pattern, this is not the case; instead the intensities increase. The detailed explanation for the increase is still a matter of debate, and, therefore, no further discussion will be presented here [Elser, 1985, 1986; Suryanarayana and Jones, 1987].

13
Dislocations

13.1. Definitions

dislocation: (1) A distortion produced by translation of the lattice by d/2, i.e., nonintegral multiple of the separation distance between two planes [Weertman and Weertman, 1964]. The resulting imperfection in the translational periodicity may be in the form of a line that lies in the plane of translation. The lattice points near the line are distorted. If the translation occurs perpendicular to the line, the dislocation is edge type. If the translation occurs parallel to the line, the dislocation is screw type. (2) A shear applied to a lattice results in slip. The boundary of the slipped region is called a shear dislocation; the displacement is **b**; the dislocation can be edge or screw type [Hirth and Lothe, 1984].

edge dislocation: A dislocation that has its Burgers vector perpendicular to the dislocation line. An edge dislocation can be produced by the addition of a half plane during a nonconservative process.

conservative motion: The number of atoms and lattice sites is unchanged by the motion [Hirth and Lothe, 1984].

climb: Motion of the dislocation out of its plane of glide [Hirth and Lothe, 1984].

plane of glide: Shear displacement produced by one dislocation [Hirth and Lothe, 1984].

slip: Glide of a number of dislocations [Hirth and Lothe, 1984].

Burgers circuit: Distance around a closed curve lying on lattice points.

Burgers vector = b: (1) Given a dislocation line, assign a unit vector **t** to the line. The sense of **t** defines the sense of the line. **b** is defined by a clockwise, closed circuit abound the line as axis in the right-handed screw sense. Redraw this circuit in a perfect lattice. The circuit will not close, the difference being defined as the Burgers vector, **b**. This convention is called FS/RH (finish-start/right hand) [Hirth and Lothe, 1984]. (2) Draw a closed circuit in a perfect crystal lattice. Redraw this circuit in the lattice containing the dislocation. The circuit now is too long by an amount **b**. This amount is defined to be the Burgers vector and the **b** so defined uses the SF/RH (start-finish/right hand) convention, and the Burgers vector so defined is called the local Burgers vector. In general, the local Burgers vector is expressed as

$$b = \int \frac{du}{dl} \, dl, \tag{13.1}$$

where **u** = elastic displacement around the dislocation, and l = line coordinate. Note that there is no universal sign convention for **b**. The Burgers vector is by definition a lattice vector since the closed circuit lies only on lattice points [Hirth and Lothe, 1984]. (3) The Burgers vector is the difference vector between the Burgers circuit that includes the dislocation and the circuit without the dislocation; the sense of the Burgers vector is positive from the end point of the circuit to the starting point [Amelinckx, 1964].

mixed dislocations: $b_{screw} = (b \cdot t) \, t$, $b_{edge} = t \times (b \times t)$; $b = b_{edge} + b_{screw}$ [Hirth and Lothe, 1984, Weertman and Weertman, 1964].

normal to the glide plane: The glide plane associated with displacement of **t** parallel to **b** is defined by **b** x **t** . For screw dislocations, any plane **p** such that **b** • **p** = 0, i.e., any plane for which **b** is a zone axis, is a possible glide plane [Hirth and Lothe, 1984].

cross slip: Movement of a dislocation from a plane **p** to a plane **q**; **p**, **q** intersect as **p** x **q** and **b** is parallel to **p** x **q** [Hirth and Lothe, 1984].

Frank energy criterion: The energy of a screw or edge dislocation is proportional to $|\mathbf{b}|^2$. For a perfect dislocation with components, $\mathbf{b} = \mathbf{b}_1 + \mathbf{b}_2$, the dislocation is unstable if $b^2 > b_1^2 + b_2^2$, since the energy of **b** is greater than the sum of the energies of its components. The result is probable dissociation of **b** into $\mathbf{b}_1$ and $\mathbf{b}_2$ [Hirth and Lothe, 1984].

imperfect dislocations: Partial dislocations = dislocations associated with stacking faults with low energy (relative to the misfit energy near the perfect dislocation core). The presence of imperfect dislocations affects twinning, phase transformations, dislocation interactions, climb, and cross slip [Hirth and Lothe, 1984].

13.2. Image Contrast of Dislocations

The image intensity or contrast of a dislocation arises from the phase change introduced by the breakdown of the lattice periodicity [Thomas and Goringe, 1979; Edington, 1979; Reimer, 1988; Nikolaichik and Khodos, 1989]. Hence, in general, there is a displacement vector, **R**, which represents the departure from periodicity. The phase can be expressed as

$$\phi = 2 \pi i \, \mathbf{g} \cdot \mathbf{R} \tag{13.2}$$

where **g** = reciprocal lattice vector. Physically the phase represents the difference in phase across the boundary introduced by the dislocation. Note that there is no phase change if **R** = 0, and that under kinematic conditions $2\pi i \, \mathbf{g} \cdot \mathbf{R} = 2\pi i \, n$, where n is an integer.

There are three simple cases of interest:

(i) $\mathbf{g} \cdot \mathbf{R} = 0$: No contrast is visible; **R** is perpendicular to **g**.

(ii) $2\pi \, \mathbf{g} \cdot \mathbf{R} = 2\pi n$, n integral. This yields a constant phase function. These conditions also apply for a fault.

(iii) $\mathbf{g} \cdot \mathbf{R} = n$, n nonintegral. The nonintegral value of n implies that **b** is not a lattice translation vector. Phase arises from a fault and **b** is the Burgers vector associated with the fault.

For any displacement, **R**, in the lattice,

$$\mathbf{R} = \Sigma \, \mathbf{R_i} = \mathbf{R_x} + \mathbf{R_y} + \mathbf{R_z}, \tag{13.3}$$

where $\mathbf{R_i}$ = components along the axes defining the beam direction and the beam is defined to be parallel to $\mathbf{R_z}$. Hence, only components perpendicular to the transmitted beam axis contribute contrast.

From the definition of dislocations, for screw dislocations, **R** is parallel to **b**. So **g** • **b** = 0 implies that the screw dislocation is not visible. For edge dislocations, **R** = $\mathbf{R_b} + \mathbf{R_n}$ = components parallel and normal to **b**. If the extra plane is parallel to the beam, then **g** • **b** = n = 0, ±1, ±2, ... Contrast may go to zero. If the extra plane is normal to the beam, then $\mathbf{g} \cdot \mathbf{b} = 0$ and $\mathbf{g} \cdot \mathbf{R_n} = m = 0, \pm 1, \pm 2, \dots$. Contrast may **not** go to zero. For mixed dislocations, the contrast conditions are a combination of the screw and edge dislocation conditions. Faulted loops have a line of no contrast, Because sections of the loop satisfy the $\mathbf{g} \cdot \mathbf{R_n} = 0$ condition. Spherical precipitates also have a line of no contrast for identical reasons. The line of no contrast is perpendicular to the reciprocal lattice vector, g, used to produce the image.

The reasons for this somewhat complex behavior of contrast can be appreciated by examining the effects on diffraction introduced by the presence of the dislocation [Nikolaichik and Khodos, 1989]. The diffraction conditions are perturbed by the disruption in local periodicity, which translates to a change in the deviation parameter, s. Hence,

$$s'_g = s_g + \mathbf{g} \cdot \frac{\partial R(z)}{\partial z}. \tag{13.4}$$

The analytic expressions for the displacement vector parallel to **b** and perpendicular to **b** are

$$\mathbf{R_b} = \frac{1}{2\pi} [\ \mathbf{b} \tan^{-1} (\frac{-z}{x}) + \mathbf{b_e} \frac{-xz}{2(1-\nu)(x^2+z^2)}, \tag{13.5}$$

and

$$\mathbf{R_n} = \frac{(\mathbf{b} \times \mathbf{u})}{2\pi(1-\nu)} \left[(1-\nu) \ln \sqrt{x^2+z^2} + \frac{x^2-z^2}{2(x^2+z^2)} \right], \tag{13.6}$$

where $\mathbf{b_e}$ is the edge component of the Burgers vector, $\mathbf{u}$ is the unit vector along the dislocation line, ν is Poisson's ratio, x is the coordinate oriented perpendicular to the dislocation and lying in the slip plane, and z is the coordinate perpendicular to the dislocation and to x (right-handed coordinates).

Taking the derivative of these functions gives

$$\mathbf{g} \cdot \frac{\partial R(z)}{\partial z} = -\frac{1}{2\pi} \left\{ \mathbf{g} \cdot \mathbf{b} \frac{x}{x^2+z^2} + \mathbf{g} \cdot \mathbf{b_e} \frac{x(x^2-z^2)}{2(1-\nu)(x^2+z^2)} \right.$$

$$\left. -\frac{\mathbf{g} \cdot \mathbf{b} \times \mathbf{u}}{1-\nu} \left[(1-2\nu) \frac{z}{x^2+z^2} - \frac{2x^2z}{(x^2+z^2)^2} \right] \right\}, \tag{13.7}$$

which clearly shows the $\mathbf{g} \cdot \mathbf{b}$ dependence is a special case of a more general set of conditions.

13.3. Analysis of Burger's Vector

Identification of the vector **b** is accomplished using the $\mathbf{g} \cdot \mathbf{b} = 0$ criterion. The magnitude and direction of **b** can be found by orienting the foil such that for two or more **g** vectors, the dislocation image becomes invisible. For cubic lattices, the analysis is aided by theoretical values of $\mathbf{g} \cdot \mathbf{b}$ for various planes. For hcp lattices this is not possible in general, but contrast using prism and basal planes (for which l = 0, removing the c/a dependence, or for which h,k,i are all zero) can be predicted. Tables 13.1-13.11 contain a listing of $\mathbf{g} \cdot \mathbf{b}$ values for various planes in FCC, BCC and hcp (for prism and basal planes). The analysis of the Burgers vector is generally complex, meaning the the use of the invisibility criterion in every case is unwise and will lead to incorrect results. See Nikolaichik and Khodos [1989] for a review of dislocation analysis.

13.4. Thompson Tetrahedron for Face-Centered Cubic

The Thompson tetrahedron is formed on 1/8 of the FCC unit cell. One corner of the tetrahedron is located at the origin. The edge length is $a\sqrt{2}/2$. Each edge contains a <110> glide direction (6 edges total); the faces contain {111} planes (4 total). See Figure 13.1.

Apexes: origin at D; others are A, B, C clockwise about the normal to the plane formed by the triangle ABC.

$\alpha,\beta,\gamma,\delta$:   midpoint of planes opposite to A, B, C, D, respectively.

a, b, c, d: planes opposite A, B, C, D, respectively, outside the tetrahedron.

a, b, c, d: planes opposite A, B, C, D, respectively, inside the tetrahedron.

Note that the notation [> and <] is used by Hirth and Lothe [1984] to indicate in a figure the sense of the vector direction. The <] and [> notation is also used to indicate permutations of the hk indices in tetragonal systems when referring to slip or dislocation movement [Hug et al., 1988].

Vector rules for the Thompson tetrahedron from Hirth and Lothe [1984] are as follows: The notation used in describing dislocations perfect and imperfect in terms of the vectors located in the Thompson tetrahedron are: **PQ** = the vector from P to Q; for vectors **PQ** and **RS**, **PQ/RS** = the vector from the midpoint of **PQ** to the midpoint of **RS**. The algebra rules for this notation are

(i) **PQ** = − **QP**,

(ii) **PD/RS** = − **RS/PQ**,

(iii) **PQ/RS** = **QP/RS** = **PQ/SR** = **QP/SR**,

(iv) **PQ** + **QR** = **PR**,

(v) **PQ** + **RS** = **PR/QS**.

In addition, $CD/\alpha\beta = \alpha\beta/\gamma\delta = \gamma\delta/\mathbf{AB}$.

Extension of the Thompson tetrahedron to hcp can be accomplished using the scheme shown in Figure 13.2.

13.5. Partials

Frank partial: b = 1/3<111> = Aα or βB: produced by the collapse of disc faces of a dislocation loop produced by a vacancy. The fault is intrinsic.

Shockley partial: b = 1/6<112>; dislocations are glissile on {111} planes.

Close packing is maintained for intrinsic, extrinsic, and twin faults. The notation A, B, C represents the three possible positions of lattice sites when viewed along the <111> direction, and the A position serves as the reference position for B and C:

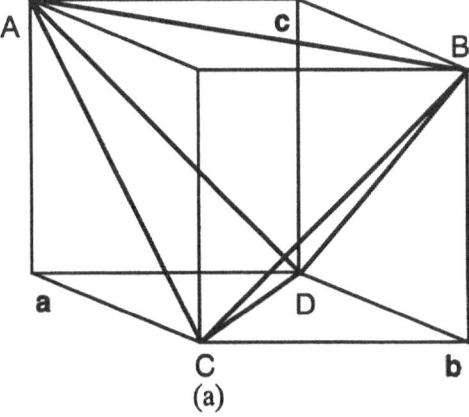

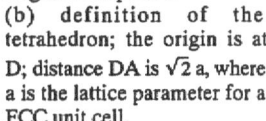

Figure 13.1. The Thompson tetrahedron. (a) the tetrahedron unfolded from the origin at D such that the observer looks toward the origin at the planes.
(b) definition of the tetrahedron; the origin is at D; distance DA is $\sqrt{2}$ a, where a is the lattice parameter for a FCC unit cell.

(a)

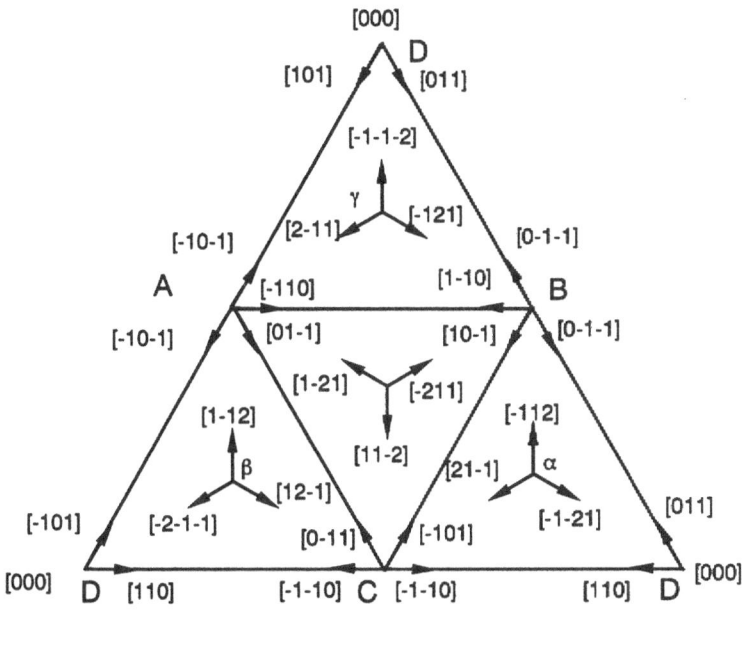

(b)

Intrinsic FCC fault = stacking change along <111>: ABCABC | BCABC; one layer is "missing"; the fault is formed by two twinning operations, one on either side of the fault.

Extrinsic FCC fault = along <111>:

ABCABC|BABCABC;

one "extra" plane; equivalent to two twin planes separated by two atomic planes.

Twin FCC stacking = along <111>:

ABCABC|BACBA.

Partials in hcp: There are two intrinsic faults and one extrinsic fault:
Intrinsic I_1:

...ABABA|ABAB... + 1/3[$\bar{1}$100] shear above the A plane before the fault
= ...ABABA|CACACA...
Intrinsic I_2:

...ABAB|CACA...   produced directly by shear of A to C and B to A.

Extrinsic E:

...ABABACBABABAB...   "extra" plane inserted.

Note that I_1 to E and E to I_1 can be accomplished by a single shear: I_1 to E: shear A to B, C to A starting with the first A after the fault. E to I_1: shear B to A, A to C starting with the first B after the fault.
    Zonal dislocations: Zonal dislocations account for fault formation, dislocation extension, and nonbasal slip in hcp. The displacements are spread over three layers, rather than one layer as in a conventional dislocation. Shuffling of the atom sites (and atoms) is required for the first and second layers upon passing of the line defect. The third layer is shifted by 1/4 the line defect vector; so if **B'C** is the defect vector in HLD notation, then **B'C** dissociates into a zonal dislocation and three partial dislocations of 1/4**B'C**. [Hirth and Lothe, 1984, pp. 362-365; Kronberg, 1961; Rosenbaum, 1964].
    Faults in BCC: Close packing in BCC arises from stacking of {112} planes as ...ABCDEFABCDEF..., where the letters refer to A as the reference layer and each layer is translated parallel to the A layer by a different vector. The types of faults

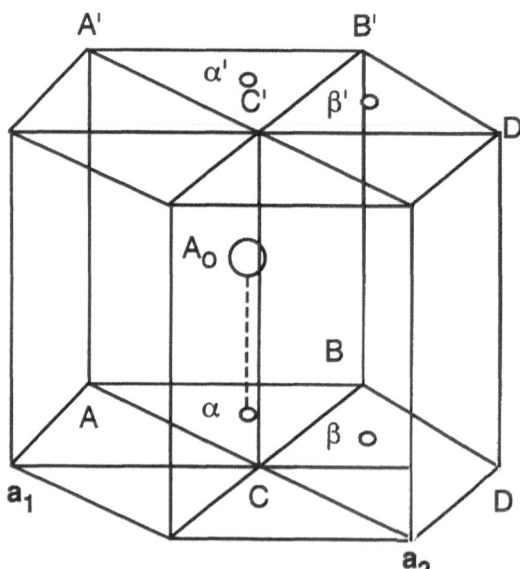

Figure 13.2. Hirth-Lothe-Damiano notation for vectors in the hcp unit cell used for defining dislocations. [Hirth and Lothe, 1984]. The α and β sites correspond to the C and B stacking sites. The plane represented by A_0 is, thus, a B plane, giving the ...ABABA... stacking of hcp.

possible are

Intrinsic I_1: ...FEDCBAFE|FEDCBA...,

Intrinsic I_2: ...FEDCBAFE|BAFEDCBA... .

Both of these faults are produced by 1/6[111] shear or by insertion of two adjacent planes.

Extrinsic E: ...CDEFABE|FCDEFABC...,

produced by the insertion of two adjacent planes, or by shear to yield the two adjacent plane pair.

Twin T: ...FEDCBAFEFABCDEF...,

produced by shear, yielding symmetric stacking about a plane.

BCC partials arise from decomposition of 1/2[111] into 1/3[111], 1/6[111], or 1/12[111]. There is a tetrahedron corresponding to the Thompson tetrahedron [Wuthrich, 1977].

13.6. Twins

Deformation of a lattice in specific ways produces a replica of the lattice whose orientation with respect to the original lattice is changed. If certain geometric conditions are satisfied, the new orientation is called a twin. Twins are defined in terms of several planes which have specific orientations with respect to one another. The reference plane for a twin is the *twinning plane*, usually symbolized by K_1. Lying in K_1 is a direction vector, η_1. Perpendicular to K_1 and containing η_1 is the *plane of shear*. A twin is present if a shear has occurred on the plane of shear such that a plane in the unsheared parent crystal has been moved through an angle of 2ϕ about an axis lying in the twin plane, the axis being perpendicular to direction vector η_1, and the movement has produced no net change in the sheared plane other than its relative orientation with respect to the twin plane. The geometry involved is shown in Figure 13.3. The sheared plane is designated K_2 and a direction vector in K_2 and in the plane of shear is designated η_2. The plane K_2 is also known as an *invariant plane*, because no change in the relative positions of lattice points or atom sites has occurred, other than the change in relative position with respect to the twin plane.   There are two simple types of twin. Type I twins are equivalent to a 180° rotation of K_2 about the normal to the twin plane, yielding a reflection of the lattice, the mirror plane being defined by the normal to the twin plane K_1 and the normal to the plane of shear. Type II twins are equivalent to a rotation of K_2 about η_1, which is equivalent to a reflection of K_1 through the twin plane K_1.

In Type I twins, K_1 and η_2 are rational, and in a Type II twin K_2 and η_1 are rational. When all four elements of the twin are rational the twin is a compound twin.

In more complex lattices or structures, the definition of twinning is more complex, especially when the structure is ordered and bonding is important [Christian, 1965]. In the general case, the number of types of twins is four: (I) reflection in K_1; (II) rotation of 180° about η_1; (III) reflection in the plane normal to η_2; (IV) rotation of 180° about the direction normal to K_1.

When the lattice only is considered, this division into four possibilities is adequate. When, however, the structure is considered, in which atom occupancy of specific sites is important, the basis for defining a twin must be extended. Bevis and Crocker [1969] and Bilby and Crocker [1965] consider this problem in some detail theoretically. The result is that an additional type is possible in which rotation about the normal to the plane of shear may occur. Since the derivation of this result is beyond the scope of this book, the reader is referred to these papers for details. When analyzing deformation twinning in ordered materials, these refinements must be considered.

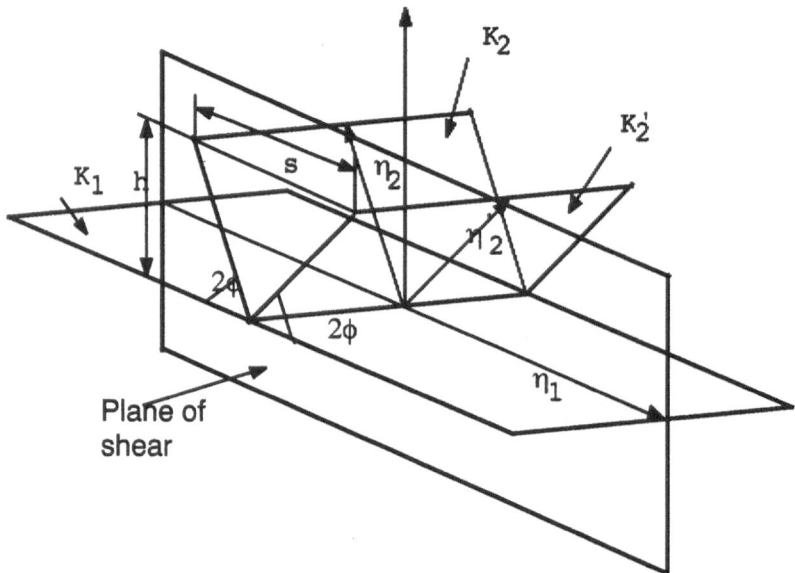

Figure 13.3. Defining geometry for twinning. The twinning plane is K_1, the twinned plane is K_2: the plane of shear is normal to K_1 and K_2. The direction η_1 lies in K_1 and the plane of shear; the direction η_2 lies in the twinned plane and the plane of shear. The shear angle is defined by ϕ; the shear distance by s at height h from the twinning plane. The twinned plane K_2 is invariant, even though the shear takes place by glide along a plane parallel to the twinning plane.

Table 13.1. Stair-Rod Reactions with Values of b^2 for Shockley Partials.

Reaction Number	Shockley partial b_2+b_3	Stair-rod product	b_1^2	$b_2^2 + b_3^2$
1	$\delta B + B\gamma$	$\delta\gamma = 1/6<110>$	2 x$a^2/36$	12 x $a^2/36$
2	$\delta B + \gamma A$	$\delta\gamma/BA =$ $1/3<001>$	4 x $a^2/36$	12 x $a^2/36$
3	$\delta C + D\gamma$	$\delta D/C\gamma =$ $1/3<110>$	8 x $a^2/36$	12 x $a^2/36$
4	$\delta B + \gamma D$	$\delta\gamma/BD =$ $1/6<013>$	10 x $a^2/36$	12 x $a^2/36$
5	$\delta B + D\gamma$	$\delta D/D\gamma =$ $1/6<123>$	14 x $a^2/36$	12 x $a^2/36$

After Hirth and Lothe [1984, Table 10-2, p. 322].

Table 13.2. Major Thompson Tetrahedron Vectors and Associated Directions.

Edge vectors	a face vectors	b face vectors	c face vectors	d face vectors
DA [101]	Bα [1$\bar{1}$2]	Aβ [$\bar{1}$12]	Aγ [$\bar{2}$1$\bar{1}$]	Aδ [$\bar{1}$2$\bar{1}$]
DB [011]	Cα [$\bar{2}$11]	Cβ [$\bar{1}\bar{2}$1]	Bγ [1$\bar{2}$1]	Bδ [2$\bar{1}\bar{1}$]
DC [110]	Dα [121	Dβ [211]	Dγ [112]	Cδ [$\bar{1}\bar{1}$2]
BA [1$\bar{1}$0]				

a face = BCD = ($\bar{1}$1$\bar{1}$); b face = ACD = ($\bar{1}\bar{1}\bar{1}$); c face = ABD = ($\bar{1}\bar{1}$1); d face = ABC = (111). The magnitude of Bα, Cα, Dα, etc., is 1/3 the magnitude of the corresponding [uvw] vector.

Table 13.3. Burgers Vectors of Stable Dislocations in FCC, BCC, hcp, Diamond Cubic and NaCl.

Lattice	Stable	Marginal	Relative Energy for Glide Plane, Anisotropic Edge Dislocation.
FCC	1/2<110>	<100>	{110}<{111}<{100}or {111}<{110}
BCC	1/2<111>, <100>	-----------	{110}<{112}<{123}
hcp	1/3<11$\bar{2}$0>, <0001>	1/3<11$\bar{2}$3>	(0002) or {1$\bar{1}$00}
Diamond cubic	1/2<110>	<100>	{110}<{111}<{100}
NaCl	<110>	<200>	{110} or {100}

After Hirth and Lothe [1984, Table 9-1].

Table 13.4. g · b for FCC Lattice.

Plane	±[110]	±[1̄10]	±[101]	±[1̄01]	±[011]	±[01̄1]
111	±1	0	±1	0	±1	0
1̄11	0	∓1	0	∓1	±1	0
11̄1	0	±1	0	±1	0	∓1
111̄	±1	0	±1	0	0	∓1
200	±1	±1	±1	±1	0	0
020	±1	∓1	0	0	±1	±1
002	0	0	±1	∓1	±1	±1
220	±2	0	±1	±1	±1	±1
22̄0	0	±2	±1	±1	∓1	∓1
202	±1	±1	±2	0	±1	±1
202̄	±1	±1	0	±2	∓1	∓1
022	±1	∓1	±1	±1	0	±2
022̄	±1	∓1	∓1	±1	±2	0

For detailed table, see Edington [1980].

Table 13.5. FCC Imperfect Dislocations; Values of g·b for Various Planes and for b = 1/6<110>.

Plane	$\frac{1}{6}[110]$	$\frac{1}{6}[101]$	$\frac{1}{6}[011]$	$\frac{1}{6}[10\bar{1}]$	$\frac{1}{6}[1\bar{1}0]$	$\frac{1}{6}[01\bar{1}]$
2 0 0	1/3	1/3	0	1/3	1/3	0
0 2 0	1/3	0	1/3	0	-1/3	1/3
0 0 2	0	1/3	1/3	-1/3	0	-1/3
2 2 0	2/3	1/3	1/3	1/3	0	1/3
2 0 2	1/3	2/3	1/3	0	1/3	-1/3
0 2 2	1/3	1/3	2/3	-1/3	-1/3	0
1 1 1	1/3	1/3	1/3	0	0	0
1 1 $\bar{1}$	1/3	0	0	1/3	1/3	0
1 $\bar{1}$ 1	1/3	1/3	0	0	1/3	1/3
$\bar{1}$ 1 1	0	0	1/3	1/3	-1/3	-1/3
1 1 $\bar{1}$	0	0	1/3	0	1/3	0
$\bar{1}$ 1 $\bar{1}$	0	0	-1/3	-1/3	1/3	0
1 1 3	1/3	2/3	2/3	0	0	-1/3
1 3 1	2/3	1/3	2/3	-1/3	-1/3	1/3
1 3 $\bar{1}$	2/3	1/3	2/3	0	-1/3	1/3
3 1 1	2/3	2/3	1/3	1/3	1/3	0
1 1 $\bar{3}$	1/3	-1/3	-1/3	2/3	0	2/3
1 $\bar{3}$ 1	-1/3	1/3	-1/3	0	2/3	-2/3
$\bar{3}$ 1 1	-1/3	-1/3	1/3	-2/3	-2/3	0
3 1 $\bar{1}$	2/3	1/3	1/3	1/3	1/3	0
1 $\bar{3}$ $\bar{1}$	2/3	0	0	2/3	-2/3	2/3
3 $\bar{1}$ 1	1/3	2/3	1/3	2/3	1/3	1/3
1 $\bar{1}$ 3	1/3	2/3	1/3	2/3	1/3	1/3
1 $\bar{3}$ $\bar{1}$	-1/3	0	-2/3	2/3	2/3	-1/3

Table 13.6. FCC Imperfect Dislocations; Values of g·b for Various Planes and for b = 1/3<111>.

Plane	$\frac{1}{3}[111]$	$\frac{1}{3}[11\bar{1}]$	$\frac{1}{3}[1\bar{1}1]$	$\frac{1}{3}[\bar{1}11]$	$\frac{1}{3}[1\bar{1}\bar{1}]$	$\frac{1}{3}[\bar{1}1\bar{1}]$	$\frac{1}{3}[\bar{1}\bar{1}1]$
2 0 0	2/3	2/3	2/3	-2/3	2/3	2/3	-2/3
0 2 0	2/3	2/3	-2/3	2/3	-2/3	2/3	-2/3
0 0 2	2/3	-2/3	2/3	2/3	-2/3	-2/3	2/3
2 2 0	4/3	4/3	0	0	0	0	-4/3
2 0 2	4/3	0	4/3	0	0	-4/3	0
0 2 2	4/3	0	0	4/3	-4/3	0	0
1 1 1	1	1/3	1/3	1/3	-1/3	-1/3	-1/3
1 1 $\bar{1}$	1/3	1	-1/3	-1/3	1/3	1/3	-1
1 $\bar{1}$ 1	1/3	-1/3	1	-1/3	1/3	-1	1/3
$\bar{1}$ 1 1	1/3	-1/3	-1/3	1	-1	1/3	1/3
1 $\bar{1}$ $\bar{1}$	-1/3	1/3	1	-1	1	1/3	1/3
1 1 3	5/3	-1/3	-1/3	1	-1	-1	1/3
1 3 1	5/3	1	-1/3	1	-1	1/3	-1
3 1 1	5/3	1	1	-1/3	1/3	-1	-1
3 1 1	5/3	5/3	-1	-1	1/3	1	-5/3
1 1 3	-1/3	-1	5/3	-1	1	-5/3	1
1 3 1	-1/3	-1	-1	5/3	1	1	1
3 1 1	-1/3	5/3	5/3	1/3	-5/3	-5/3	1
1 1 3	1	1/3	-1	1/3	-1/3	1	-5/3
3 1 1	1	5/3	5/3	-1	-1/3	-1/3	-5/3
3 1 1	1	1/3	1/3	-1	-5/3	1	-1/3
1 1 3	-1	1	5/3	-5/3	-5/3	-1/3	-1
1 3 1	-1	-1/3	-1/3	-5/3	5/3	-5/3	1/3

Table 13.7. FCC Imperfect Dislocations; Values of g·b for Various Planes and for b = 1/6<112>.

Plane	$\frac{1}{6}[112]$	$\frac{1}{6}[1\bar{2}1]$	$\frac{1}{6}[\bar{2}11]$	$\frac{1}{6}[211]$	$\frac{1}{6}[\bar{1}12]$	$\frac{1}{6}[1\bar{2}1]$	$\frac{1}{6}[21\bar{1}]$	$\frac{1}{6}[\bar{2}11]$	$\frac{1}{6}[\bar{1}\bar{2}1]$	$\frac{1}{6}[11\bar{2}]$	$\frac{1}{6}[\bar{1}12]$	$\frac{1}{6}[2\bar{1}\bar{1}]$
200	1/3	1/3	2/3	1/3	-2/3	1/3	2/3	-2/3	1/3	1/3	-1/3	2/3
020	1/3	2/3	1/3	-2/3	1/3	-2/3	1/3	-1/3	-1/3	-2/3	2/3	-1/3
002	2/3	1/3	1	1/3	1/3	-1/3	1/3	1/3	-1/3	2/3	-1/3	-1/3
220	2/3	1	1	-1/3	0	-1/3	1	0	-1/3	0	-2/3	1/3
202	1	2/3	1	2/3	-1/3	0	1/3	-1/3	1/3	1/3	1/3	1/3
022	1	1	2/3	-1/3	-1	-1/3	2/3	-1/3	1/3	1/3	-2/3	-2/3
111	2/3	2/3	2/3	2/3	0	0	0	0	0	0	0	0
$\bar{1}10$	-1/3	2/3	0	0	-1/3	1/3	1/3	-1/3	-1/3	-1/3	-1/3	2/3
$\bar{1}\bar{1}0$	-1/3	-1/3	-1/3	-2/3	2/3	-1/3	-1/3	-1/3	2/3	-1/3	2/3	-1/3
$\bar{1}\bar{1}1$	0	0	-1/3	1/3	0	1/3	0	0	-1/3	0	0	-1/3
$1\bar{1}0$	-2/3	-1/3	-2/3	1/3	1/3	1/3	-2/3	1/3	2/3	2/3	2/3	-1/3
$\bar{1}10$	-1/3	-2/3	-1/3	1/3	1/3	2/3	-1/3	1/3	-2/3	-1/3	-1/3	-1/3
113	4/3	1	1	1/3	0	0	0	0	0	0	0	0
131	1	4/3	1	1/3	1	2/3	2/3	-1	-1	2/3	-1/3	2/3
311	1	1	1	-2/3	-1	-1	1	1/3	1/3	-1/3	-1/3	2/3
110	-2/3	0	0	0	1/3	1	-1/3	1	-1	1	-4/3	2/3
$\bar{1}31$	0	-2/3	-1/3	0	-1	-1	1	1/3	1/3	-1/3	2/3	2/3
$\bar{3}11$	0	0	1/3	1/3	1/3	1	-1/3	1	-1	1	2/3	-4/3
$1\bar{1}3$	1	1/3	1	1	-1	-1	1	-1	1	-1	1	0
130	1/3	1	2/3	0	0	-2/3	1/3	1/3	1	1/3	-1	0
310	1/3	2/3	1/3	-1	2/3	1/3	4/3	-2/3	-1/3	2/3	0	1
$3\bar{1}1$	2/3	1/3	1	-1	1/3	2/3	2/3	-4/3	0	4/3	-1/3	1
$\bar{1}\bar{1}0$	-1	-2/3	-1/3	-1	4/3	1	0	0	-1	0	0	1
$\bar{1}30$	-2/3	-1	-1/3	1	-2/3	4/3	2/3	-2/3	-2/3	1/3	-1	-1
$3\bar{1}0$	-1/3	-1/3	-2/3	1	-1/3	-2/3	-1/3	-1/3	-4/3	0	0	-1
$\bar{3}\bar{1}1$	2/3	0	-1	1	-1/3	-1/3	-2/3	1	2/3	-2/3	1	0
$1\bar{1}3$	0	2/3	-1/3	1	-1	-1/3	4/3	-1/3	-4/3	1/3	0	-2/3
$\bar{1}30$	0	2/3	-2/3	2/3	-2/3	-1	-1	1	-2/3	4/3	-2/3	-2/3
$3\bar{1}0$	0	0	2/3	-4/3	2/3	1/3	1/3	-1	-2/3	-2/3	4/3	4/3

Table 13.8. g·b Values for BCC Perfect Dislocations.

Plane	$\pm[111]$	$\pm[11\bar{1}]$	$\pm[1\bar{1}1]$	$\pm[\bar{1}11]$	$\pm[100]$	$\pm[010]$	$\pm[001]$
200	±1	±1	±1	∓1	±1	0	0
$\bar{2}$00	∓1	∓1	±1	±1	∓1	0	0
020	±1	±1	∓1	±1	0	±1	0
0$\bar{2}$0	∓1	∓1	±1	∓1	0	∓1	0
002	±1	∓1	±1	±1	0	0	±1
00$\bar{2}$	∓1	±1	∓1	∓1	0	0	∓1
110	±1	±1	0	0	±1/2	±1/2	0
1$\bar{1}$0	0	0	±1	∓1	±1/2	∓1/2	0
011	±1	0	0	±1	0	±1/2	±1/2
01$\bar{1}$	0	±1	∓1	0	0	±1/2	∓1/2
101	±1	0	±1	0	±1/2	0	±1/2
10$\bar{1}$	0	±1	0	∓1	±1/2	0	∓1/2
112	±2	0	±1	±1	±1/2	±1/2	±1
121	±2	±1	0	±1	±1/2	±1	±1/2
211	±2	±1	±1	0	±1	±1/2	±1/2
11$\bar{2}$	0	±2	∓1	∓1	±1/2	±1/2	∓1
1$\bar{2}$1	0	±2	±2	∓1	±1/2	∓1	±1/2
$\bar{2}$11	0	±2	∓1	±2	∓1	±1/2	±1/2
1$\bar{1}$2	∓1	±1	0	∓2	±1/2	∓1/2	∓1
12$\bar{1}$	∓1	0	±1	∓2	±1/2	∓1	∓1/2
$\bar{2}$1$\bar{1}$	∓1	0	∓2	±1	∓1	±1/2	∓1/2
$\bar{1}$1$\bar{2}$	∓1	±1	∓2	0	∓1/2	±1/2	∓1
$\bar{1}$21	∓1	∓2	±1	0	∓1/2	∓1	±1/2
$\bar{2}$$\bar{1}$1	∓1	∓2	0	±1	±1	∓1/2	±1/2

Table 13.9. Details of Various Vectors Used in hcp Vector Notation for Dislocations.

b or glide plane notation	vector notation	(4,4) or Miller-Bravais notation
Perfect dislocation Burgers vector	**AC**	$\frac{1}{3}[1\bar{2}10]$ (along a_2)
	AA'	$[0001]$ (along c)
	CB'	$\frac{1}{3}[\bar{1}\bar{1}23]$ (along a+c)
Glissile partials		
(equivalent to Shockley partials)	**Aa**	$\frac{1}{3}[\bar{1}100]$
	Ab	$\frac{2}{3}[\bar{1}100]$
Sessile partials		
(equivalent to Frank partials)	**aA$_0$**	$\frac{1}{2}[0001]$
	Ab'	$\frac{1}{3}[\bar{2}203] = \frac{1}{3}[\bar{2}200] + [0001]$
	BA$_0$	$\frac{1}{6}[20\bar{2}3] = \frac{1}{6}[20\bar{2}0] + \frac{1}{2}[0001]$
	Ca'	$\frac{1}{3}[0\bar{1}13] = \frac{1}{3}[0\bar{1}10] + [0001]$
Partials on $\{10\bar{1}0\}$	**eC**	$\frac{1}{18}[4\bar{2}\bar{6}3] = \frac{1}{18}[4\bar{2}\bar{6}0] + \frac{1}{6}[0001]$
Glide planes	**ABC**	(0001)
	ABB'A'	$(0\bar{1}10)$
	AB'C'	$(1\bar{1}01)$
	CA$_0$B'	$(11\bar{2}2)$

After Hirth and Lothe [1984, Table 11-11].

Table 13.10. Hcp Perfect Dislocations; Values of g·b for Various Planes and for b = 1/3<$\bar{1}\bar{1}$23> and 1/3[0003].

Plane	$\frac{1}{3}[\bar{1}\bar{1}23]$	$\frac{1}{3}[\bar{1}2\bar{1}3]$	$\frac{1}{3}[2\bar{1}\bar{1}3]$	$\frac{1}{3}[0003]$
1 $\bar{1}$ 0 1	1	0	2	1
$\bar{1}$ 1 0 1	1	2	0	1
0 $\bar{1}$ 1 1	0	2	1	1
0 $\bar{1}$ 1 1	2	0	1	1
$\bar{1}$ 0 1 1	0	1	2	1
$\bar{1}$ 0 0 1	2	1	0	1
0 0 0 1	1	1	1	1
0 0 0 1	$\bar{1}$	2	2	1
1 $\bar{1}$ 2 1	2	$\bar{1}$	$\bar{1}$	1
$\bar{1}$ 2 $\bar{1}$ 1	2	2	0	1
1 $\bar{1}$ 1 1	3	0	0	1
2 $\bar{1}$ $\bar{1}$ 1	0	3	3	1
1 $\bar{1}$ 2 1	0	2	2	1
$\bar{1}$ 2 $\bar{1}$ 1	$\bar{2}$	2	$\bar{2}$	1
3 $\bar{1}$ 2 1	3	3	$\bar{2}$	1
3 $\bar{1}$ 2 1	3	$\bar{2}$	3	1
2 $\bar{1}$ 3 1	2	3	3	1
2 $\bar{1}$ 3 1	2	$\bar{2}$	0	1
$\bar{1}$ 3 $\bar{2}$ 1	$\bar{2}$	2	0	1
1 3 $\bar{1}$ 1	4	$\bar{1}$	4	1
3 1 $\bar{2}$ 1	$\bar{1}$	4	4	1
3 2 $\bar{1}$ 1	$\bar{1}$	0	$\bar{1}$	1
2 3 $\bar{1}$ 1	0	$\bar{1}$	4	1
2 1 $\bar{3}$ 1	4	0	$\bar{1}$	1

Table 13.11. hcp Perfect Dislocations; Values of g·b for Various Planes and for b = 1/3<$\bar{1}\bar{1}$20>.

Plane		$\frac{1}{3}[2\bar{1}\bar{1}0]$	$\frac{1}{3}[\bar{1}2\bar{1}0]$	$\frac{1}{3}[\bar{1}\bar{1}20]$
1 $\bar{1}$ 0	n	1	$\bar{1}$	0
$\bar{1}$ 1 0	n	$\bar{1}$	1	0
0 1 $\bar{1}$	n	0	1	$\bar{1}$
0 $\bar{1}$ 1	n	0	$\bar{1}$	1
1 0 $\bar{1}$	n	1	0	$\bar{1}$
1 0 0	n	$\bar{1}$	0	1
0 0 0	n	0	0	0
1 $\bar{1}$ 2	n	1	1	$\bar{2}$
$\bar{1}$ 2 $\bar{1}$	n	$\bar{2}$	1	1
1 $\bar{1}$ 1	n	$\bar{1}$	$\bar{1}$	2
2 $\bar{1}$ $\bar{1}$	n	$\bar{1}$	2	$\bar{1}$
1 $\bar{1}$ 2	n	2	$\bar{1}$	$\bar{1}$
$\bar{1}$ 3 $\bar{2}$	n	1	2	$\bar{3}$
3 $\bar{1}$ 2	n	$\bar{3}$	$\bar{1}$	2
2 $\bar{1}$ $\bar{1}$	n	$\bar{3}$	2	2
2 3 $\bar{1}$	n	2	$\bar{3}$	1
2 $\bar{1}$ $\bar{3}$	n	2	1	$\bar{3}$
1 $\bar{1}$ 2	n	$\bar{1}$	$\bar{2}$	3
3 $\bar{1}$ 3	n	$\bar{1}$	3	$\bar{2}$
3 2 $\bar{1}$	n	3	$\bar{1}$	$\bar{2}$
2 3 $\bar{1}$	n	$\bar{2}$	$\bar{2}$	$\bar{1}$
2 1 $\bar{3}$	n	$\bar{2}$	$\bar{1}$	3

References

Amelinckx, S., *The Direct Observation of Dislocations*, No. 6 in Solid State Physics series, eds. F. Seitz and D. Turnbull, Academic Press, New York, 1964.

Ashcroft, N. W. and N. D. Mermin, *Solid State Physics*, Saunders College, Philadelphia, 1976.

Azaroff, L., *Elements of X-Ray Crystallography*, McGraw-Hill, New York, 1968.

Bancel, P., P. A. Heiney, P. W. Stephens, A. I. Goldman, and P. M. Horn, *Phys. Rev. Lett.*, **54**(22), 2422-2425 (1985).

Barrett, C. S., "Crystal Structure of Metals," in *Metals Handbook*, Vol. 8, 8th edition, ASM International, Metals Park, OH, 1973, pp. 233-241.

Barrett, C. S. and T. B. Massalski, *Structure of Metals*, Pergamon Press, New York, 1980.

Beers, Y., *Introduction to the Theory of Error*, Addison-Wesley, Reading, MA, 1957.

Bevis, M. and A. G. Crocker, *Proc. Roy. Soc. Lond. A*, **315**, 509-529 (1969).

Bilby, B. A. and A. G. Crocker, *Proc. Roy. Soc. Lond. A*, **288**, 240-255 (1965).

Brown, P. J. and J. B. Forsyth, *The Crystal Structure of Solids*, Edward Arnold, London, 1979.

Buerger, M. J. *X-Ray Crystallography*, John Wiley and Sons, New York, 1962.

Buxton, B. F., J. A. Eades, J. W Steeds, and G. Rackham, *Phil. Trans.*, **281**, 171-194 (1976): "The Symmetry of Electron Diffraction Zone Axis Patterns."

Cahn, J. W., D. Shechtman, and D. Gratias, *J. Mater. Res.*, **1**, 13-26 (1986).

Christian, J. W., *The Theory of Transformations of Metals and Alloys*, Pergamon Press, London, 1965.

Cullity, B.D., *Elements of X-Ray Diffraction*, Second Edition, Addison-Wesley Publishing Co., Reading, MA, 1978.

Cundy, H. M. and A. P. Rollett, *Mathematical Models*, 1961 or other editions.

Dahmen, U., *Acta Met.*, **30**, 63-73 (1982).

Deer, W. A., R. A. Howie, and J. Zussman, *An Introduction to the Rock-Forming Minerals*, Longman Group, Ltd., London, 1980 [available through Halsted Press, Division of John Wiley & Sons, Inc., New York].

Edington, J. W., *Practical Transmission Electron Microscopy*, Philips, Mahwah, NJ, 1980.

Elser, V., *Phys. Rev. B*, **32**(8), 4892-4898 (1985): "Indexing Problems in Quasicrystal Diffraction."

Elser, V., *Acta Cryst.*, A42, 36-43 (1986):"The Diffraction Pattern of Projected Structures."

Gjonnes, J. and A. Moodie, *Acta Cryst.*, **19**, 65-67 (1965).

Goldstein, J. and H. Yakowitz (Eds.), *Practical Scanning Electron Microscopy*, Plenum Press, New York, 1975.

Hahn, T., Ed., *International Tables for Crystallography*, Vol A, D. Reidel, Boston, 1983.

Hirth , J. and J. Lothe, *Dislocations,* John Wiley and Sons, New York, 1984.

Hug, G., A. Loiseau , and P. Veyssiere, *Phil. Mag. A*, **57**(3), 499-523 (1988).

Hull, D., *Introduction to Dislocations*, Pergamon Press, New York, 1975.

ITC, *International Tables for Crystallography*, Vol A, T. Hahn (Ed.), D. Reidel, Boston, 1983.

ITXC, *International Tables for X-Ray Crystallography*, Vols. I, II, III, IV, Kathleen Lonsdale (Ed.), The Kynoch Press, Birmingham, UK, 1959.

Jackson, A. G., *J. of Elect. Microsc. Tech.* **5**, 373-377 (1987): "Prediction of holz Pattern Shifts in Convergent Beam Diffraction."

Jackson, A. G., *Ultramicroscopy*, **30**, 349-354 (1989).

Jackson, A. G., *Ultramicroscopy*, **32**, 181-182 (1990).

Jackson, A. G. and M. Saqib, private communication, 1988.

Katz, A. and M. Duneau, *Scripta Met.*, **20**, 1211-1216 (1986): "Quasiperiodic Tilings obtained by Projection."

Kitaigorodskii, A. I., *Organic Chemical Crystallography*, Consultants Bureau, New York, 1955.

Kittel, C., *Introduction to Solid State Physics*, John Wiley and Sons, 3rd Edition or later, 1967.

Kronberg, M. L., *Acta Met.*, **9**, 970 (1961).

Ladd, M. F. C. and R. A. Palmer, *Structure Determination by X-Ray Crystallography*, Plenum Press, New York, 1988.

Machlin, E. S., *Acta Met.* **22**, 95-108 (1974): "Pair Potential Model of Intermetallic Phases - I", and *Acta Met.* **22**, 109-121 (1974): "Pair Potential Model of Intermetallic Phases - II."

Massalski, T. B. (Ed.), *Binary Alloy Phase Diagrams*, Vol. 2, ASMI, Metals Park, OH, 1986.

Miller, I. and J. E. Freund, *Probability and Statistics for Engineers*, 2nd Edition, Prentice-Hall, Englewood Cliffs, N. J., 1977 (or later editions).

Nabarro, F. R. N., *Theory of Crystal Dislocations*, Dover Publications, New York, 1987.

Nelson, David R., "Quasicrystals," *Sci. Amer.*, **255**, 45 (1986).

Nikolaichik, V. I. and I. I. Khodos, *J. of Micros.*, **155**, part 2, 123-167 (1989): "A Review of the Determination of Dislocation Parameters using Strong- and Weak-Beam Electron Microscopy."

Otte, H. M. and A. G. Crocker, *Phys. Stat. Sol.* **9**, 441 (1965): "Crystallographic Formulae for Hexagonal Lattices."

Parthe, E., *Crystal Chemistry of Tetrahedral Structures*, Gordon and Breach, New York, 1964.

Partridge, P..G., *Met. Rev.*, **12**, 169 (1967): "The Crystallography and Deformation Modes of Hexagonal Close-Packed Metals."

Pearson, W. B., *The Crystal Chemistry and Physics of Metals and Alloys*, Wiley-Interscience, New York, 1972.

Pettifor, D. G. , *Sol. State Comm.*, **51**(1), 31-34 (1984): "A Chemical Scale for Crystal-Structure Maps."

Pettifor, D. G., *New Scientist*, May 29, 1986: "New Alloys from the Quantum Engineer."

Pettifor, D. G., *J. Less Comm. Mets.*, **114**, 7-15 (1985): "Phenomenological and Microscopic Theories of Structural Stability."

Pettifor, D. G., *J. Phys. C*: Solid State Phys., **19**, 285-313 (1986): "The Structures of Binary Compounds: I. Phenomenological Structure Maps."

Pettifor, D. G. and R. Podloucky, *J. Phys. C: Solid State Phys.*, **19**, 315-330 (1986): "The Structures of Binary Compounds: II. Theory of the pd-Bonded AB Compounds."

Pettifor, D. G. and R. Podloucky, *Phys. Rev. Lett.*, **55**(2), 261 (1985): response to comment on letter in *Phys. Rev. Lett.*

Raghavan, M., J. C. Scanlon, and J. W. Steeds, *Met. Trans.*, **15A**, 1299 (1984): "Use of Reciprocal Lattice Layer Spacing in Convergent Beam Electron Diffraction Analysis."

Reimer, L., *Transmission Electron Microscopy*, Springer-Verlag, New York, 1988.

Rosenbaum, H. S., *Deformation Twinning*, P. E. Reed-Hill, J. P. Hirth, and H. C. Rogers (Eds.), Gordon and Breach Science Publishers, New York, 1964, p. 43.

Shechtman, Dan, *Scr. Met.*, 20, 1185-1186 (1986): "Introduction to Papers on Crystals with Fivefold Symmetry."

Stoter, L. P., *Norelco Reporter*, **31**, No. 2EM, Sep. 1984, p. 36-44.

Suryanarayana, C. and H. Jones, *Int. J. of Rapid Sol.*, **3**, 253-293 (1987): "Formation and Characteristics of Quasicrystalline Phases: A Review."

Tanaka, M., H. Sekii, and T. Nagasawa, *Acta Cryst.*, **A39**, 825-837 (1983): "Space-Group Determination by Dynamic Extinction in Convergent-Beam Electron Diffraction."

Tanaka, M., H. Sekii, and T. Nagasawa, *Acta Cryst.*, **A40**, 721 (1984): "Space-Group Determination by Dynamic Extinction in Convergent-Beam Electron Diffraction: Errata."

Tanaka, M. and M. Terauchi, *Convergent Beam Electron Diffraction*, JEOL Ltd, Tokyo, 1985.

Tanaka, M., M. Terauchi, and T. Kaneyama, *Convergent Beam Electron Diffraction II*, JEOL Ltd, Tokyo, 1988.

Tanner, L. E., and W. J. Leamy, "The Microstructure of Order-Disorder Transitions", in *Order-Disorder Transformations in Alloys*, Ed.: H. Warlimont, Springer-Verlag, New York, 1974, p. 180.

Thomas, G., and M. J. Goringe, *Transmission Electron Microscopy of Materials*, John Wiley & Sons, New York, 1979.

Vainstein, B. K., A. A. Chernov and L. A. Shuvalov (editors), *Modern Crystallography*, Volumes I-IV, Springer-Verlag, New York, 1979.

Verma, A. R., and O. N. Srivastava, *Crystallography for Solid State Physics*, John Wiley and Sons, New York, 1981.

Villars, P., and L. D. Calvert, *Pearson's Handbook of Crystallographic Data for Intermetallic Phases*, Vols 1, 2, 3, ASM International, Metals Park, OH, 1985.

Wahlstrom, E. E., *Optical Crystallography*, 5th Edition, John Wiley and Sons, New York, 1983, p. 12-19.

Weertman, J. W. and J. Weertman, *Elementary Dislocation Theory*, MacMillan, New York, 1964.

Williams, D. B. *Practical Analytical Electron Microscopy in Materials Science*, Philips Electronics Instruments, Inc., Electron Optics Publishing Group, Mahwah, New Jersey, 1983.

Wuthrich, C., *Phil. Mag.*, **35**, 337 (1977).

Index

1-fold inversion, 105
2-beam conditions, 20
2-fold axis, 102
2-fold rotation-inversion, 105
2-fold symmetry axes, 48
3-fold axis, 102
3 axis, 3, index , 4
(3, 3), 57, 64, 77
(4, 4), 64
4-fold symmetry axis, 44
4 axis, 4, index, 4
5-fold symmetries, 176
14 Bravais lattices, 101

A, B, or C sites, 28
A$_2$, 136
A$_2$B$_2$, 136
accuracy, 19
adjacent plane sequence, 33
adjacent plane spacing, 31, 32
Al, 23
allowed and extincted beams, 13, 15
allowed and extincted indices, 14
allowed planes, 10
Amelinckx, 183
analysis of patterns, 20
angle, α, between trigonal axes, 58
angle between d and R, 89
angle between directions, 39, 42, 45, 49, 54
angle between planes, 7, 24, 39, 42, 45, 49, 54, 60, 71
angle between two directions, 60,
angles between axes in reciprocal space, 6
angles in direct space, 6, 92
angular separation of lines in the disc, 136
APB, 86

approaches to identifying the planes, 21
Ashcroft and Mermin, 8, 12
atom coordinates, 32
atom scattering amplitude, 24
atom sites, 5, 189
atom species, 5
atomic number, 162
atomic weights, 162
atoms per unit cell, 42, 43, 50, 55
axes, 1
azimuthal angle, 93 95

B face-centering (b axis unique), 78
B2, 136
back focal plane, 127
Bain, 80
Bancel et al., 179
barrel distortion, 18
Barrett and Massalski, 33, 104
basal plane, 83
base-centered, 15
base-centered monoclinic, 55
base-centered orthorhombic lattices, 50
base-centered tetragonal, 46
basis vectors, 38, 41, 57
basis vectors for nonprimitive monoclinic, 79
basis, 1, 105
basis coordinate system, 2
basis set of vectors, 176
basis vector, 1, 2, 5, 8, 10, 13, 14, 16, 61, 78, 179
BCC, 24, 25, 26, 37, 43, 80, 186
BCC cube faces, 77
BCC defining primitive vectors, 12
BCC lattice parameter, 75
BCC lattices, 13
BCC partials, 189
BCC to hcp transformation, 75-77

BCC to orthorhombic transformation, 75, 76
BCT, 10
Beers, 19
Bevis and Crocker, 190
Bloch waves, 167
body-centered basis, 13
body-centered cubic, 29, 30, 32, 33, 83
body-centered orthorhombic, 50
body-centered tetragonal, 10
body-centered vectors, 12
body centering, 106
body diagonal, 33, 35, 41, 77
Bragg, 18, 19, 129, 131, 133, 134, 136
Bragg equation, 133
Bravais lattices, 28, 100, 104
Bright field, 135, 136
Brown and Forsyth, 107
Buerger, 14, 15
Burgers circuit, 183
Burgers vector, 82, 183, 186
Burgers vectors of dislocations in hcp structures, 88

Cahn, 176
camera constant equation approximation, 18
camera constant, 17, 19, 22, 24
camera length, 18
Cartesian coordinate system, 6
CBED, 127, 128, 135
center of a unit cell, 92
center of inversion, 105
center of symmetry, 102
centering, 104
centers of beams, 17
centers of three coplanar atoms, 30
Cesium, 25
chemical scale, 162
Christian, 189
classical crystallography, 106
climb, 183, 184
close-packed directions, 83
close-packed plane, 31, 32, 33, 83
close packing, 28, 31, 186
close packing in BCC, 188
close packing in three dimensions, 30

closest atom centers, 8
condenser lens, 127
condenser lens stigmator, 18
confidence level of the measurement, 20
conservative motion, 183
construction of the holz pattern, 141
contrast conditions, 185
contrast using prism and basal planes, 185
conventions used in analyzing and labeling a diffraction pattern, 21
convergence angle, 127
convergent beam, 131
convergent beam electron diffraction, 127
convergent beam patterns, 17
conversion equations, 61
conversion formulas, 75
convolution, 169
coordinate system, 57
coordination, 33, 35
cross product, 2
cross products among the trigonal basis vectors, 59
cross products in (3, 3), 68
cross slip, 184
crystal, 1, 3, 23, 75
crystal designations, 100
crystal direct lattice, 5
crystal lattice, 22, 106
crystal plane, 93
crystal potential, 170
crystal structure, 5, 89, 100
crystal structures and Fourier transforms, 170
crystal system, 1, 17, 22, 24, 25, 26, 80, 93, 100, 107
crystallographic axes, 1, 105
crystallographic point, 1
CSF, 86
cube body diagonal, 30
cube diagonal, 9, 29
cube face diagonal, 10
cubic, 4, 28, 40, 96, 107
cubic case, 91
cubic lattices, 185
cubic system, 22, 23, 41, 90, 92, 93
Cullity, 1, 96
Cundy and Rollett, 175

cyclic, 105

Dahmen, 80
dark field, 135
defining angles between basis vectors, 38
defining geometry of the N-S and the W-
    E axes, 97
defining vectors, 8, 38, 41, 44, 48, 52, 57
defining vectors for hcp structures, 14
definition of the ABC positions, 28
deflation, 177
deformation of a lattice, 189
deviation parameter s, 134-136, 185
diagonal glide, 1141
diamond, 25
difference in phase across the boundary
    introduced by the dislocation, 184
different centerings, 24
diffracted beams, 23
diffracted waves, 167
diffraction conditions, 20
diffraction pattern, 1, 17, 19, 20, 21, 26,
    90, 92, 127
diffraction pattern analysis, 17
diffraction patterns from icosahedral
    structures, 181
dihedral, 106
Dirac delta function, 169
direct lattice, 3, 5, 6, 17, 45, 49, 89, 92
direct lattice angles, 6, 54
direct lattice direction vector, 40
direct lattice projections, 89
direct space, 56
direct space lattice, 5
direction, 39
direction indices, 3
direction in the direct lattice, 53
direction vector, 1, 40
directions in the direct lattice, 3
dislocation interactions, 184
disordered to ordered transformations, 81
displacement vector, 184, 185
disruption in local periodicity, 185
distance between equivalent planes, 39,
    42, 45
distance between equivalent planes in the
    direct lattice, 49, 54, 70, 71
distance in the direct lattice, 18

distance of closest approach, 41
distortions in patterns, 18
dot product, 1, 41, 44, 48, 53, 58, 59, 64,
    67, 68, 75, 77, 78, 177, 179

edge component of the Burgers vector,
    185
edge dislocation, 183, 184
edge length of the rhombohedral cells,
    176
Edington, 184
effects on diffraction introduced by the
    presence of the dislocation, 185
elastic displacement around the
    dislocation, 183
electron density, 170
electron diffraction, 1, 18
electron diffraction pattern, 176
electron scattering function, 170
electron wavelength, 18
Elser, 176, 182
Elser indices, 176
emulsion, 17
emulsion side, 20
end-centered, 15
end-centered basis vectors, 15
end-centered lattice primitive basis
    vectors, 15
energy of a screw or edge dislocation,
    184
equal planes, 4
equations relating the direction indices
    and the plane indices for
    rhombohedral indices, 63
equator, 93
equatorial plane, 93, 95
equivalent planes, 6, 21, 34
equivalent plane sequence, 34
equivalent points, 100, 105,
error, 19
errors in measurements, 17
eucentric position, 18
Ewald, 129, 130
Ewald sphere, 1, 129, 131, 133, 134,
    135, 142
exact Bragg condition, 131, 136
extincted, 10
extinction distance, 136

extinctions, 16
extra plane is normal to the beam, 185
extra plane is parallel to the beam, 190
extrinsic E, 188, 189
extrinsic FCC fault, 187

face-centered cubic, 8, 29, 32, 33, 83
face-centered orthorhombic, 50
face-centered tetragonal, 46
face centering, 106
face centers, 33
face diagonal, 41
face diagonal of the cubic lattice, 10
face diagonals, 29
fault, 184
fault intrinsic, 186
faulted loops, 185
faults in BCC, 188
FCC, 10, 24, 25, 26, 32, 37, 80, 89, 186
FCC basis, 10
FCC unit cell, 186
FCC zolz, folz, 134
FCT basis vectors, 77
FCT to BCT and BCT to FCT
    transformation, 77
Fibonacci numbers, 175
film plate, 17
first order Laue zone (folz), 92, 130, 144
folz indices, 137
folz pattern, 92
folz ring, 135
Four axis, four index, 64
Fourier series and transforms, 167
Fourier transform, 24, 177
Fourier transform of the density function,
    170
Fourier transform of the function f(x),
    168
Frank energy criterion, 184
FS/RH (finish-start/right hand), 183

G, 130, 131, 134, 135
g • b = 0 criterion, 189
general equations applicable to any
    system, 38
general transformations, 11

geometry of points in the stereographic
    projection, 92
geometry of projections of latitudes at
    angle α and longitude at angle β, 95
geometry of the optical axis, 18
Gjonnes and Moodie, 136
glide plane, 100, 106, 136, 184
glide plane relationships, 106
glissile, 186
GM lines, 136
golden mean, 175, 179
golden ratio, 175
golden section, 175
Goldstein and Yakowitz, 20
grain, 22
great circle, 93-95
grid projections, 96

H, 131, 134
harmonic functions, 167
hcp, 25, 26, 28, 37, 77, 80, 186
hcp basis, 57
hcp CBED ring patterns, 147
hcp lattices, 191
hcp reference basis, 79
hcp reference basis vectors, 14
hcp system, 25
hcp to orthorhombic, 75, 76
hcp to rhombohedral, 61
Hermann-Mauguin symbol, 106
hexagonal basis, 57
hexagonal basis vectors, 14, 59
hexagonal basis vector reference system,
    14, 15
hexagonal close packed, 14, 30, 33, 83
hexagonal coordinate system, 14
hexagonal indices, 61
hexagonal reference system, 57
hexagonal system, 64
hexagonal vectors, 76
hierarchy of the crystal families, 100
higher order Laue zone lines, 131
higher order Laue zones (holz), 91, 129,
130, 133, 134, 135
Hirth and Lothe, 183, 186, 188
(hkl) for a specific set of integers or
    planes, 3
holz lines, 135

holz disc, 133, 135
holz radius, 133
holz rings have a radius, 143
Hug, 86

icosahedral, 179
icosahedral group, 106
icosahedron, 175
ideal hcp ratio, 33
ideal packing, 31
image contrast of dislocations, 184
image features, 20
image plane, 18
imperfect dislocations, 184
independent slip systems, 83
indexing conventions, 21
indexing diffraction patterns, 23
indexing holz Patterns, 137
indices for various hexagonal indexing
    systems, 73
indices of directions in the three indexing
    systems, 69
indices of planes, 10
indices of planes in the three indexing
    systems, 69
inflation, 177
inscribed decagon, 1175
intensity, 1
intermediate image plane, 19
intermediate lens, 18
international symbols, 106
interplanar separations, 6, 18, 22
interstitial positions, 33
interstitial sites, 33, 35
interstitial sites in the BCC structure, 35
interstitial sites in the FCC structure, 35
intrinsic FCC fault, 187
intrinsic I1, 188, 189
intrinsic I2, 188, 189
invariant plane, 189
inverse N-W, 80
inversion center, 102
isostructural transformations, 80
ITC, 105, 106
ITXC, 24, 25, 107, 167

Jackson and Saqib, 148

Jagodzinski-Wyckoff notation, 107

Katz and Duneau,179
kinematic conditions, 24, 184
kinematically forbidden lines, 136
Kittel, 30
Kronberg, 188
Kurdjumov-Sachs, 80

Ladd and Palmer, 105
latitudes, 96
lattice, 1, 5, 6, 28, 41, 43, 46, 50, 55,
    75, 81, 106, 128, 179
lattice basis vectors, 3
lattice centered on the A, B, or C face,
    106
lattice centering, 75
lattice constant(s), 1
lattice of points, 24
lattice parameter, 7, 22, 38, 41, 44, 48,
    52, 57, 75, 81, 127, 130
lattice point, 5, 24, 92, 183, 189
lattice point coordinates, 50, 55
lattice points/cell, 30
lattice sites, 32, 34, 186
lattice vector, 183
lattice vector from the origin in the zolz
    to some (hkl)L lattice point in zone L,
    91
Laue, 129
Laue class 1
Laue zone, 1, 42, 46, 49, 55, 129, 130,
    143
Laue zone number, 137
lens stigmators, 18
line coordinate, 183
line of no contrast, 185
lines of longitude, 96
local Burgers vector, 183
long form of the international symbols,
    107
long period unit, 22

m, 135
Machlin classification of some
    intermetallics, 162

macroscopic symmetry elements, 105
magnitude and direction of b, 185
magnitude of reciprocal lattice vectors,
    39, 41, 71, 74, 184
magnitudes of the hexagonal lattice
    parameters in terms of the trigonal, 58
magnitudes of the hcp vectors in terms of
    the rhombohedral magnitudes, 79
magnitude of the trigonal a, b, c vectors,
    58
magnitude of the trigonal reciprocal
    lattice vectors, 59
measured angles, 17
measured distances, 17, 23, 24, 25
Mendeleev Number, 162
Miller-Bravais, 64
Miller and Freund, 20
Miller indices, 2, 3, 64
minimum number of specific planes, 23
minimum standard deviation, 20
mirror, 105, 135
mirror plane, 102, 106
misfit energy near the perfect dislocation
    core, 184
mixed dislocations, 185
mixed dot products, 40
mixed product, 2
modified Schlafli symbol, 163
momentum wavevector, 7
monoclinic basis vectors for various
    centerings, 80
monoclinic nonprimitive to primitive
    transformation, 78
monoclinic, 4, 40, 107
monoclinic system, 54
monoclinic system (b axis unique), 52
motif, 1, 5, 90
multiple patterns, 22

Nagasawa, 136
nearest neighbor, 30
Nelson, 175
neostructural transformations, 80
net of regular hexagons, 163
net of regular triangles, 163
Nikolaichik and Khodos, 184, 185, 186
Nishijima-Wasserman, 80
nonorthonormal axes, 4

nonprimitive basis vectors, 78
nonprimitive cube, 8
normal distribution, 20
normal to the direction [uvw], 93
normal to the glide plane, 184
normal to the plane, 40
normal vectors, 92
notation [> and <], 186
notation for direction and plane indices,
    86
number of atoms in the unit cell
    Pearson, 104
number of independent slip systems for
    FCC, BCC, and hcp, 87
number of independent slip systems for
    hcp, 87
number of spheres per unit cell, 28
number of the Laue zone, 1

objective aperture, 18
objective lens, 19, 127
obverse setting, 57
octahedral, 37, 106
octahedral planes, 83
octahedral sites, 33, 35, 36
omega phase, 80
optic axis, 18
optimum contrast, 20
ordered structures, 80
orthogonal basis, 1
orthogonal coordinate system, 76
orthogonal right-handed coordinate
    system, 41, 44, 48, 52
orthohexagonal, 4, 64, 96
orthonormal, 6
orthonormal reference coordinate
    system,   1
orthorhombic, 4, 40, 107
orthorhombic basis vectors, 76
orthorhombic system, 76
orthorhombic vectors, 75

P, 1
packing fraction, 30
packing fraction and stacking sequences,
    28
pair potential model, 162

parallelogram, 163
parity table, 145, 146
Parthe, 33, 107
partials in hcp, 188
Partridge, 83
pattern, 17
pattern intensities, 128
Pearson, 33, 104, 107, 164
Pearson symbols, 104
Penrose tiling, 163
perfect dislocation, 184
periodic table, 162
permutation groups, 1106
permutation of indices, 4
permutations, 4
phase change, 184
phase transformations, 184
photometer measurements, 17
Pitsch, 80
Pitsch-Schrader, 80
plane, 2, 20, 25, 39, 83
plane indices, 2, 10, 14, 16, 38, 41, 44, 52
plane in reciprocal space, 39, 41, 45, 49, 54, 60
plane of glide, 105-106, 195
plane of shear, 189
plane projection, 93
planes in the folz, 92
planes that are equivalent, 4
point group, 100, 105, 127, 128, 135, 136
point symmetry, 1
point symmetry groups, 105
points in the trigonal cell, 57
Poisson distribution, 20
polar vector, 176
pole, 93
pole projection, 93
position vectors, 5
potentials involved in the electron scattering, 167
potentials of the crystal array, 170
precision, 19
primitive basis, 13
primitive basis vectors, 15
primitive cell, 1, 16, 105
primitive lattice, 106
primitive trigonal, 57
primitive vectors, 9, 12

principal axes, 9
prismatic directions, 83
prism type I planes, 83
prism type II planes, 83
probability that the error is less than the estimate, 20
procedure for indexing the rings, 22
projection along the [001] axis of the hcp system, 74
projection of an atom position onto P, 90
projection of a structure onto a reference plane, 90
projection of atom positions, 90
projection of the Ewald sphere, 22
projections of reciprocal lattice points, 90
projector lens, 18
protractor, 17
pyramid I, 87
pyramid II, 87
pyramidal directions, 83
pyramidal planes, 83

quasilattice constant, 17
quasiperiodicity in three dimensions, 175

Ramsdell notation, 108
ratio method, 23
ratio tables, 25
ratio of experimentally measured reciprocal lattice vector distances, 26
ratios, 23
ratios of radii, 22
ratios of segments of a line, 175
reciprocal lattice, 1, 6, 17, 18, 129, 133, 176
reciprocal lattice angles, 6, 38, 54
reciprocal lattice dot products, 53
reciprocal lattice layers, 90
reciprocal lattice points, 1, 129
reciprocal lattice separation of Laue zones, 91
reciprocal lattice vector, 17, 38, 41, 44, 48, 52, 59, 66
reciprocal lattice vectors referenced to the rhombohedral parameters, 59
reciprocal space, 14, 54, 178

reciprocal space projections, 90
reference great circle, 95
reference hcp system, 76
reference plane for a twin, 189
reflection, 105
regular plane tessellations, 163
Reimer, 18, 184
relationship between indices for planes
    and for directions, 42
relationships among the direction indices
    and the plane indices, 45, 49, 54
relationships among the indices, 72
relationships among the plane indices, 61
relationships between direction indices
    and plane indices for the three
    indexing systems, 70
relationships between the plane indices
    (h,k,l) and the direction indices
    [u,v,w], 4
repeat distance in reciprocal space, 71
repeat distances, 1
repeat separation, 4
repeat spacing, 40
repeat spacing in reciprocal space, 38,
    42, 46, 50, 55
repeat spacing in the direct lattice, 43
review of dislocation analysis, 186
rhombohedral, 4, 14
rhombohedral axes, 106
rhombohedral bases, 13
rhombohedral basis vectors, 12
rhombohedral basis vectors derived from
    BCC, 13
rhombohedral defining axes for FCC, 9
rhombohedral indices, 13
rhombohedral primitive cell, 10
rhombohedral reference basis, 79
rhombohedral to hcp, 61, 79
rhombohedral unit cell for FCC, 9
rhombus, 163
right-handed coordinate system, 2, 20
ring 17, 23, 34, 129, 130
ring diameters, 22
ring pattern from an FCC structure, 21
ring patterns, 21
rings in convergent beam diffraction, 143
Rosenbaum, 188
rotational axes, 105
rotation groups, 106

rotation-inversion, 105

SAD (selected area diffraction) aperture,
    18
SC, 25
scattering by each atom, 24
scattering function, 170
Schlafli symbols, 163
Schoenflies symbol, 106
screw axes, 100, 102
screw dislocations, 184
screw symmetry element, 106
screw type, 183
second nearest neighbor, 30
second order Laue zone, 130
Sekii, 136
selected area diffraction patterns, 90
semi-regular nets, 163
semi-regular tessellations, 163
SESF, 86
SF/RH (start-finish/right hand), 183
shape factor, 24
shift of the focal plane, 19
shift vector, 31
Shockley partial, 186
short form of the international symbol,
    107
simple cell, 1
simple cubic, 16, 28, 29, 31, 32
simple cubic basis vectors, 16
simple cubic lattice, 41
simple cubic plane projection, 34
simple cubic projection, 36
simple lattice, 106
simple monoclinic, 55
simple orthorhombic, 50
simple tetragonal, 46
simple types of twin, 189
SISF, 86
sites in a unit cell, 33
slip, 183
slip direction, 83
slip plane, 83
slip systems, 83
slip vectors, 83
Sodium, 25
solz, 130
some ordered structures, 80

some orientation relationships, 80
space group, 25, 100, 106, 127, 128
space group symbols, 105
space lattice, 5, 100
spacing between equivalent planes, 39
special transformations for BCC, hcp, and orthorhombic lattices, 75
species, 5
spherical aberrations, 18, 19
spherical precipitates, 185
spot diffraction pattern, 21-22
square wave, 167
stacking, 31-33
stacking change along <111>, 187
stacking fault acronyms, 86
stacking of layers, 28
stacking of the adjacent layers, 89
stacking sequences, 31
standard deviation, 19, 20
statistics and error, 19
stereographic projection, 92
stereographic projection conventions, 93
Stoter, 136
structure, 23, 28, 89
structure factor, 24, 177
structure of a domain, 22
structure symbols, 100
structures, 8, 22
Strukturbericht symbols, 104
Student's t distribution, 20
sum rules, 10
superdislocations, 86
Suryanarayana and Jones, 182
symmetric stacking about a plane, 189
symmetries present in two dimensional tilings of simple polygons, 163
symmetry, 5, 127, 128, 135, 136
symmetry element, 105-106
symmetry elements of the principal directions for the crystal system, 106
symmetry elements on a space lattice, 106
symmetry of the unit cell, 100
symmetry operations, 100
symmetry symbols, 105
systematic error, 19
systematic peaks, 178

table for FCC ratios, 23
tables of plane ratios, 25
Tanaka, 127, 136
Terauchi, 127
tetragonal, 4, 40, 107
tetragonal system, 25, 26, 186
tetrahedra, 175
tetrahedral, 37, 106
tetrahedral sites, 33, 35, 36
tetrahedron, 186
thickness, 128, 131, 136
Thomas and Goringe, 19, 184
Thompson tetrahedron for face-centered cubic, 186
three axis, three index, 64
three-axis, three-index basis, 57
three-dimensional Penrose tile, 176
transformation equations for BCT to FCC, 10
transformation equations for FCC to BCT, 11
transformation from trigonal to hcp vectors, 14
transformation matrix, 61
translational repeat distance, 5
transmitted beam, 22, 23
triacontahedron, 177
triclinic, 4, 40, 107
triclinic system, 56
trigonal reciprocal lattice vectors, 59
trigonal, 14, 40, 106, 107
trigonal basis vectors, 14, 57
trigonal coordinate system, 14
trigonal indices, 15, 60
trigonal system, 57
trigonal vectors, 15
triple hexagonal cell, obverse setting, 79
triple hexagonal unit cell and the primitive rhombohedral cell, 79
true value, 19
twin FCC stacking, 188
twin T, 189
twin plane, 189
twinning, 184
twinning plane, 189
twins, 189
two-dimensional projection, 31
type I twins, 189
type II twins, 189

type of lattice, 106

unit cell, 1, 29, 105, 106
unit cell parameters, 30
unit impulse function, 169
unit vector coordinate system, 15
unit vectors i, j, k, 2
universal ratios, 25

vacancy, 186
Vainstein, 1, 106, 107
value of the golden mean, 175
variations in stacking, 107
vector defining the direction D
    perpendicular to the plane P, 89
vector from the origin to a point on P
    where the projection of d terminates,
    89
vector in the direct lattice, 176
vector operations, 1
vector perpendicular to zolz and parallel
    to u, 91
vector rules for the Thompson
    tetrahedron, 186
vector projection of d onto D, 92
vector to the atom position with
    coordinates [u,v,w], 89
vertical rotation axis, 105
Villars and Calvert, 104

volume, 1, 7, 28, 29, 41, 44, 48, 53, 67
volume in direct space, 38
volume in reciprocal space, 38
volume in the hexagonal reference, 58
volume in the trigonal reference, 58
volume of a unit cell, 2

wavevector, 18, 24
Weertman and Weertman, 183
Whole field, 135, 136
Wulff net, 93

x-ray diffraction, 22

zero order Laue zone, 39, 91, 129
Zhdanov notation, 108
Zincblende, 25
zolz, 91, 92, 129, 130, 135
zolz disc, 133
zolz indices, 137
zonal dislocations, 188
zone axes, 1, 17, 40, 127, 130, 135, 184
zone axis method, 24
zone law, 92
zone law for higher order Laue zones,
    42, 45, 49, 54
zone law for zero order Laue zone, 49,
    54